Springer

Tokyo
Berlin
Heidelberg
New York
Barcelona
Hong Kong
London
Milan
Paris
Singapore

S. Kashii, A. Akaike, Y. Honda (Eds.)

Nitric Oxide in the Eye

With 99 Figures, Including 3 in Color

 Springer

Satoshi Kashii, M.D., Ph.D.
Associate Professor
Department of Ophthalmology and Visual Sciences
Graduate School of Medicine, Kyoto University
54 Kawahara-cho, Shogoin, Sakyo-ku, Kyoto 606-8507, Japan

Akinori Akaike, Ph.D.
Professor and Chairman
Department of Pharmacology
Graduate School of Pharmaceutical Sciences, Kyoto University
46-29 Shimoadachi-cho, Yoshida, Sakyo-ku, Kyoto 606-8501, Japan

Yoshihito Honda, M.D., Ph.D.
Professor and Chairman
Department of Ophthalmology and Visual Sciences
Graduate School of Medicine, Kyoto University
54 Kawahara-cho, Shogoin, Sakyo-ku, Kyoto 606-8507, Japan

ISBN-13: 978-4-431-68017-8   Springer-Verlag Tokyo Berlin Heidelberg New York

Library of Congress Cataloging-in-Publication Data

Nitric oxide in the eye / S. Kashii, A. Akaike, Y. Honda (eds.).
    p. ; cm.
    "The symposium was held in Kyoto, Japan on September 28 and 29 as a satellite
symposium of the XII International Congress of Eye Research, 1996"—Pref.
    Includes bibliographical references and index.
    ISBN-13: 978-4-431-68017-8   e-ISBN-13: 978-4-431-67949-3
    DOI:10.1007/978-4-431-67949-3
    1. Nitric oxide—Physiological effect—Congresses. 2. Nitric
oxide—Pathophysiology—Congresses. 3. Eye—Physiology—Congresses. 4.
Eye—Pathophysiology—Congresses. I. Kashii, S. (Satoshi), 1952– II. Akaike, A.
(Akinori), 1951– III. Honda, Yoshihito, 1939– IV. International Congress of Eye
Research (12th : 1996 : Kyoto, Japan)
    [DNLM: 1. Eye Diseases—drug therapy—Congresses. 2. Nitric Oxide—therapeutic
use—Congresses. WW 166 N731 2000]
    QP535.N1 N5527 2000
    612.8′4—dc21
                                                                        00-039483
Printed on acid-free paper

© Springer-Verlag Tokyo 2000
Softcover reprint of the hardcover 1st edition 2000

Typesetting, printing, and binding: Best-set Typesetter Ltd., Hong Kong
SPIN: 10763731

Preface

Nitric oxide (NO) is a simple gas with free radical properties. No one would have imagined a role for such a simple substance in the human body. In 1998, R.F. Furchgott, F. Murad, and L.J. Ignarro received the Nobel prize for their work on NO. Interestingly, Alfred B. Nobel, who invented dynamite by combining nitroglycerin with other substances, took nitroglycerin for his chest pain without realizing that NO mediates its action. Now, in addition to its vasodilating action, NO is known to possess many fundamental functions that include neurotransmission, blood pressure control, blood clotting, and immune responses. These diverse functions, conversely, imply that the simple NO molecule may unite neuroscience, physiology, and immunology and may change our understanding of how cells communicate and defend themselves. In this context, the International Symposium, Nitric Oxide and Free Radicals, was organized to address current thinking about the widespread distribution and variety of functions of NO in the eye. The symposium was held in Kyoto, Japan, September 28–29 as a Satellite Symposium of the XII International Congress of Eye Research, 1996. The Symposium was the first international gathering of leading scientists and ophthalmologists meeting to present and discuss their most recent results in a specialized area of research, specifically concerning the eye. Taking advantage of this opportunity, we planned to publish a monograph about NO in the eye, with the aim of creating an up-to-date review for ophthalmologists as well as for research scientists who are interested in this rapidly progressing field. Most of the contributors were participants in the symposium. We asked them, as leaders in the field, to provide a broad overview of their subject and to relate their discussion to functional aspects and clinical implications where possible.

This monograph aims to describe the current knowledge on the widespread distribution and variety of functions of NO in the eye. Many novel properties and actions of NO in the eye have been reported in recent years. NO is an important intracellular messenger in the regulation of aqueous humor dynamics. Activation of the NO/cGMP system results in relaxation of the trabecular meshwork and increases aqueous humor outflow and decreases intraocular pressure. Numerous ocular autonomic nerves contain nNOS and likely use NO as a neurotransmitter. These nitrergic (nitroxidergic) nerves are involved in regulating the blood supply to the eyes of most mammalian species. The activation of nitrergic nerves results in relaxation of the central retinal arterial smooth muscles. In humans, the choroid is under the influence of a basal release of NO, which maintains the vasodilatory tone of choroidal vessels. In primates, where light is focused on the fovea, the uniquely dense nitrergic vasodilative innervation in this specific area is important for a reflexive increase in the choroidal blood flow and is thought to be involved in the process of accommodation. The iris sphincter is innervated by cholinergic nerves. NO inhibits the cholinergic contraction of the sphincter through the formation of cGMP and dilates the pupil. NO could be an ideal mydriatic eyedrop that can enlarge the pupil quickly in a short time, with rapid reversion to normal size. In a pathological state, such as uveitis or retinal ischemia, NO is known to mediate the cytotoxic process. However, NO alone seems unable to induce cytotoxicity. Peroxynitrite formed by the combination of NO˙ (radical form) and superoxide anions leads to retinal toxicity. In retinal ischemia, overstimulation of the NMDA subtype of glutamate receptor has been shown to be responsible for tissue injury. As suggested by Lipton, the ideal NO group donor drug is one that can react readily with the critical thiol group of the redox modulatory site on the NMDA receptor to down-regulate the receptor's activity, and that will offer neuroprotection against glutamate toxicity such as in retinal ischemia.

The reader will soon recognize that this monograph is the first to be published that covers all areas of the eye in relation to NO. Monographs published earlier have concentrated largely on the pharmacology of NO in the central and peripheral nervous systems or cardiovascular systems. This book summarizes much of our current understanding of NO in the eye. It is intended not only for those already involved in the field but also for ophthalmologists who need a comprehensive, up-to-date text on the subject. We hope that our work will result in increased interest in the fascinating and important subject of NO in the eye.

SATOSHI KASHII
AKINORI AKAIKE
YOSHIHITO HONDA

Contents

IV. The Uvea: Pathophysiology of Uveitis

V. The Retina

Contributors

I. Overview

Nitric Oxide: Some New Concepts

THOM MITTAG

Introduction

There has been an explosive growth of research on the role of nitric oxide (NO) in physiological and pathological processes in many tissues and organs. The eye shares in this phenomenal growth. Rather than give an overview of the various systems in the eye where NO plays a role, I discuss herein some new concepts relating to NO function emerging from other studies that may have future relevance to the eye. The topics covered are listed in Table 1.

Nitric Oxide Synthase Localization

Although several constitutive isoforms of nitric oxide synthase (NOS) are known, research suggests that in some tissues the localization of NOS is important to its function. Table 2 lists the major tissues where localization has been studied. To date, the only known localization mechanisms are binding (anchoring) of NOS by fatty acid amide formation at the amino-terminus (NOS III) and by dystrophin in striated muscle membranes (NOS I). In one paper (Brenman et al., 1995) it was shown that in patients with Duchenne-type muscular dystrophy, where dystrophin is defective or absent, there is no staining for NOS I in striated muscle membranes (Fig. 1). It is not known whether the absence of NOS contributes to the muscle pathology in these patients or even what the role of NOS is at this locus in normal muscle. Dystrophin is also expressed in the retina at a level much higher than in other neural tissues and is prominently localized in the outer plexiform layer (Fig. 2). The retinas of muscular dystrophy patients show a loss of dystrophin at this locus, where NADPH-diaphorase activity (which indicates the presence of NOS) is also prominently localized (Pillers et al., 1993; Yamamoto et al., 1993). It is interesting and perhaps not just coincidental that the retina, as shown in Fig. 3,

Department of Ophthalmology, Mount Sinai School of Medicine, 1 Gustave L. Levy Place, New York, NY 10029, USA

TABLE 1. Nitric oxide: new concepts

| NOS isoforms and localization |
| Secondary signaling: nitrosothiols |
| Hemodynamics |
| Protection from NO toxicity |

NOS, nitric oxide synthase; NO, nitric oxide.

TABLE 2. NOS localization

Isozyme	Anchor	Tissue/form
eNOS (III)	Consensus N-myristoylation site at amino-terminus	Vascular endothelial cells/soluble and particulate fractions
nNOS (I)	?	CNS and PNS/sol and particulate
nNOS (I)	?	Lung epithelial cells
nNOS (I)	?	Kidney
nNOS (I)	Dystrophin	Striated muscle cells/particulate

e, epithelial; n, neuronal; CNS, central nervous system; PNS, peripheral nervous system.

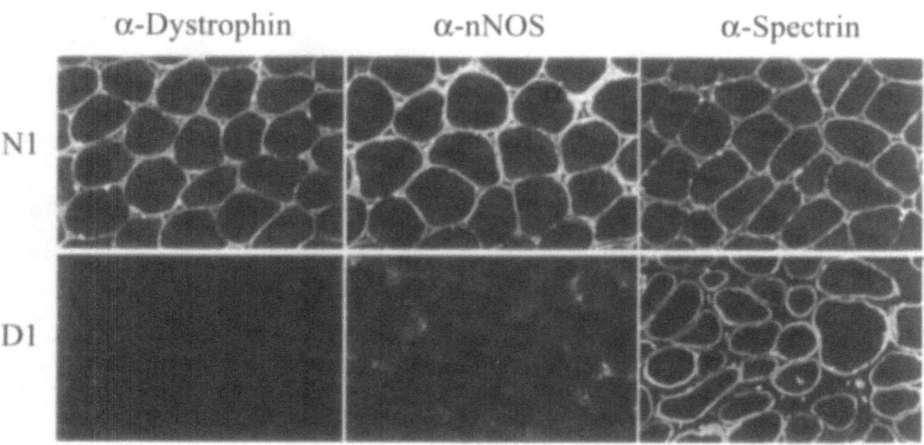

FIG. 1. Cryosections of normal (*N1*) and Duchenne muscular dystrophic (*D1*) skeletal muscles, immunostained (α) for dystrophin, neuronal nitric oxide synthase (nNOS), and spectrin. There is an absence of dystrophin and NOS I in dystrophic muscle

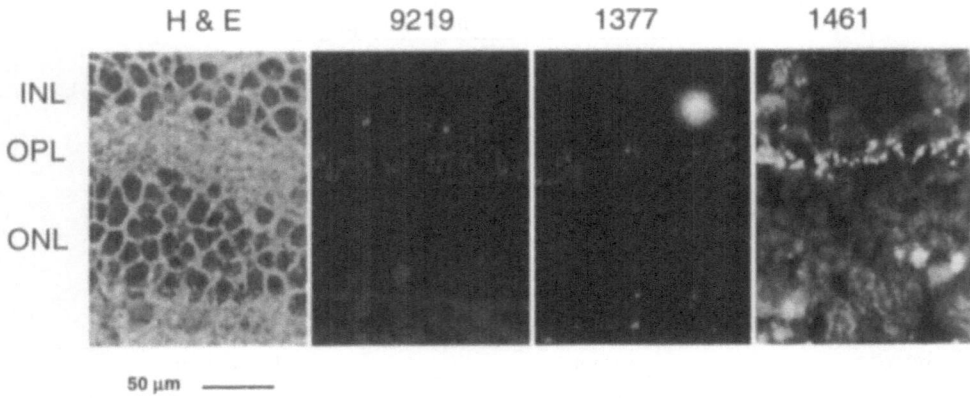

FIG. 2. Immunohistochemical staining of cryosections of adult human retina showing dystrophin staining, particularly by antiserum no. 1461, in the outer plexiform layer (*OPL*) and isolated cells in the outer nuclear layer (*ONL*). H + E, hematoxylin and eosin-stained section

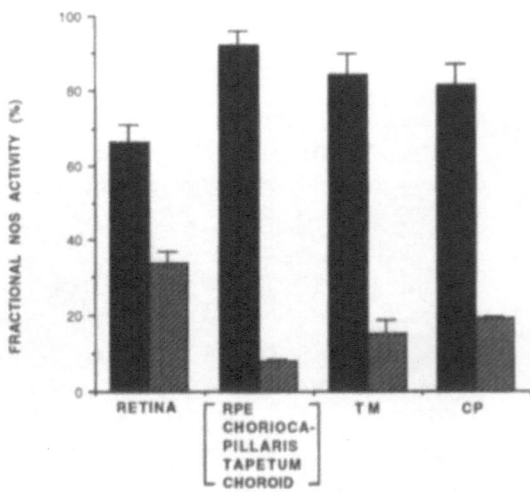

FIG. 3. Fractional activity of cytosolic (*solid bars*) and particulate (*hatched bars*) NOS activity in tissues dissected from the bovine eye. *RPE*, retinal pigment epithelium; *TM*, trabecular meshwork; *CP*, ciliary processes. (Data from Geyer, Podos, and Mittag)

has the highest proportion of particulate-bound NOS activity of all the ocular tissues. Some of this fraction may be bound to dystrophin. In muscular dystrophy patients the scotopic electroretinogrophic (ERG) b-wave is greatly diminished, whereas the photopic ERG responses are almost normal (Pillers et al., 1993). These findings indicate that there may be a defect in rod electrophysiological responses in muscular dystrophy subjects, which suggests that dystrophin, NOS, or both may be involved in the electrophysiological coupling of rod photoreceptor cells and Müller cells, where the b-wave originates.

Physiologic Signaling by NO

It is likely that NO can function as a signaling system via nitrosothiols. It is of course well known that NO can act as an intracellular signal when complexed to the bound iron atom in the guanylyl cyclase enzyme, thereby activating it and generating cyclic guanosine monophosphate (cGMP). However, it has been shown that complexes of NO with heme and nitrosothiols, which are secondary products of NO, may be important extracellular signals in hemodynamics (Jia et al., 1996). Before discussing the possible involvement of these NO-derived messengers in the circulation, it is necessary to discuss briefly the chemistry of NO in a physiological setting. The nitrosation of thiols to nitrosothiols (RSNO) can proceed via two reactions, for simplicity termed "uncatalyzed" and "catalyzed," though these descriptions are not strictly true. The reactions are summarized in Fig. 4. The uncatalyzed reaction is the result of several chemical steps, of which the most important rate-limiting step is the initial reaction of NO with dissolved O_2 (not shown). Careful measurements of this reaction indicate that it is much too slow to account for nitrosothiol formation, and therefore it is much more likely that in a physiological situation RSNOs are formed by the "catalyzed" reaction (e.g., by Fe ions, as shown in Fig. 4). The Fe involved in RSNO formation in vivo is most likely protein-bound Fe, as in hemoglobin (Hb). It should also be noted that exchange of the sulfur-bound NO between thiols (the transnitrosation reaction in Fig. 4) readily takes place. The equilibrium position of this reaction depends on the conditions (e.g., pH) and nature of the different thiol structures (denoted by R and R*). The interactions of NO with hemoglobin, which contains both a bound Fe and a reactive thiol group that is readily nitrosolated (cysteine 93 of the β-chain, or cys β 93, in human hemoglobin) further illustrates that these reactions may have physiological relevance. Figure 5 (top) shows the reactions

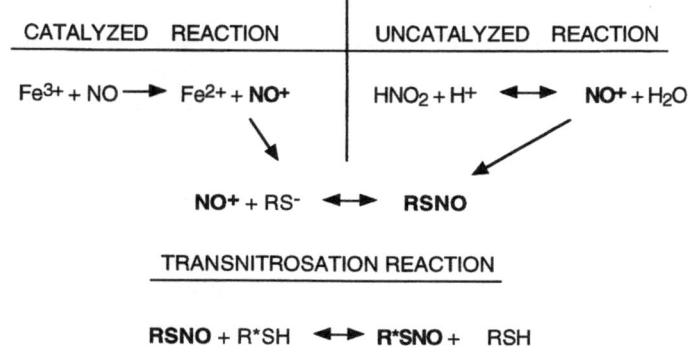

Fig. 4. Chemical reactions for formation of nitrosothiols (RSNO) from thiols and for nitroso exchange (transnitrosation) among various thiols (RSH, R*SH)

REACTIONS OF NO WITH HEME OF Hb

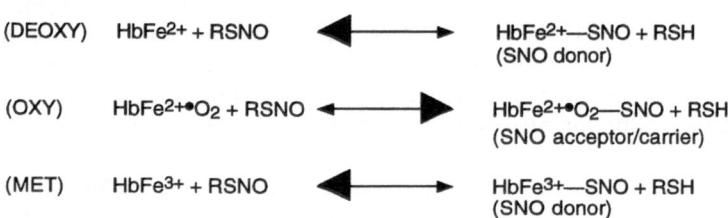

(DEOXY) $HbFe^{2+} + NO$ $\xrightarrow[\text{instantaneous}]{}$ $HbFe^{2+} \bullet NO$ (eqn 1)

(OXY) $HbFe^{2+} \bullet O_2 + NO$ $\xrightarrow{}$ $HbFe^{3+} + NO_3^-$ (eqn 2)

(MET) $HbFe^{3+}—SH + NO$ $\xrightarrow{}$ $HbFe^{2+}— SNO + H^+$ (eqn 3)

TRANSNITROSATIONS OF RSNO WITH CYSβ93 OF Hb

(DEOXY) $HbFe^{2+} + RSNO$ ⬅➤ $HbFe^{2+}—SNO + RSH$ (SNO donor)

(OXY) $HbFe^{2+}\bullet O_2 + RSNO$ ⬅➤ $HbFe^{2+}\bullet O_2—SNO + RSH$ (SNO acceptor/carrier)

(MET) $HbFe^{3+} + RSNO$ ⬅➤ $HbFe^{3+}—SNO + RSH$ (SNO donor)

FIG. 5. (*Top*) Reactions of NO with the heme group of hemoglobin (*Hb*) in its deoxy, oxy, and met forms. (*Bottom*) Transnitrosation equilibria between *RSNO*, for example *S*-nitrosoglutathione, and cysteine β 93 of hemoglobin in its deoxy, oxy, and met forms

of NO with the heme group of hemoglobin in its various forms. It has long been known that deoxyhemoglobin is a rapid scavenger of NO (Fig. 5, eqn 1) and that oxyhemoglobin is oxidized by NO to methemoglobin (Fig. 5, eqn 2). However, the ferric ion in methemoglobin has the potential to function as the "catalyst" for the formation of the SNO group in a reaction analogous to the "catalyzed" reaction shown in Fig. 4, resulting in the nitrosation of cys β 93 (Fig. 5, eqn 3).

The formation of SNO in hemoglobin (Hb-SNO) has been studied in more detail using nitrosothiols of endogenous molecules, glutathione SNO and *S*-nitrosocysteine (Jia et al., 1996). The SNO of cys β 93 in hemoglobin can participate in transnitrosation equilibria in the presence of such physiologically relevant thiols. Although these reactions are difficult to study, evidence suggests that the equilibria differ for the three forms of Hb-SNO (deoxy-, oxy-, and met-hemoglobin) as shown in Fig. 5 (bottom). For deoxy- and methemoglobin the equilibria lie more to the left, as indicated by the larger arrowheads; hence these forms of Hb-SNO can function as SNO donors in transnitrosation reactions; with the oxy-Hb-SNO form the equilibrium lies more to the right, and it can function as an SNO acceptor/carrier.

The main question now is whether these interactions of hemoglobin and NO have physiological or pharmacological significance. First, can these reactions occur in vivo with regard to NO complexing to the heme group? Because NO is highly diffusible it seems likely that at the sites where NO is formed in close contact with red blood cells (e.g., by endothelial cells in the arterial microcirculation and lung capillaries) reactions of NO with hemoglobin occur with the forms of hemoglobin locally present in accordance with the reactions shown in Fig. 5 (top). With regard to nitrosation of cys β 93 in the three forms of hemoglobin, they would occur inside red blood cells either by the intramolecular reaction of methemoglobin with NO (Fig. 5, eqn 3) or by transnitrosation from low-molecular-weight nitrosothiols formed separately from reaction with NO, such as S-nitrosoglutathione.

The second question is whether the biological responses to Hb and Hb-SNOs are in accordance with the predictions for the various reaction products of hemoglobin interacting with NO. Table 3 shows the results reported by Jia et al. (1996) on blood pressure responses to hemoglobin derivatives injected into the femoral vein of the rat. Oxyhemoglobin, an NO scavenger, caused the expected hypertension, but, this was not observed with injection of the corresponding cys β 93 SNO derivative. Thus the NO scavenging reaction of the heme group (eqn 1, Fig. 5) in oxy-Hb-SNO appears to be prevented by the presence of the NO group on cys β 93 via an intramolecular neighboring group effect. Another possibility is that there is physiological antagonism; that is, the vasoconstriction caused when NO is removed by the heme scavenging reaction is opposed by a transnitrosation from the cys β 93 SNO causing relaxation, both occurring in the same hemoglobin molecule. Relaxation could also occur when the injected compound ($HbFe^{2+} \cdot O_2$-SNO), an SNO carrier, gives up its O_2, becoming $HbFe^{2+}$-SNO, which is an SNO donor (Fig. 5, bottom). This interpretation is perhaps favored by the strong hypotensive response to injection of met Hb-SNO which is also an SNO donor (Table 3).

The final question is whether the various species of hemoglobin interacting with NO have physiological relevance. Table 4 contains data obtained and recalculated from the paper of Jia et al. (1996), who analyzed rat blood from venous and arterial circulations for the concentration of hemoglobin heme complexes with NO (Hb·NO) and for their Hb-SNO content. Table 4 shows that their sum approximates 1 µM and is essentially the same in circulating venous and arterial blood. However, there is a major difference in the proportion of the two forms: Only 3% is in the form of Hb-SNO in venous blood, whereas it comprises

TABLE 3. Blood pressure response to Hb and Hb-SNO species[a]

Hb, Hb-SNO species injected	Blood pressure response (mmHg)
$HbFe^{2+} \cdot O_2$ (NO scavenger)	$+20 \pm 3$
$HbFe^{2+} \cdot O_2$—SNO (SNO carrier)	-3.0 ± 2.5
$HbFe^{3+}$—SNO (SNO donor)	-34 ± 5

Hb-SNO, hemoglobin-cysteine β nitrosothiol.
[a] After 0.2 µmol/kg was injected into the femoral vein.

TABLE 4. Content and distribution of Hb·NO and Hb-SNO in rat blood in vivo

Blood site	Hb·NO + Hb-SNO (nM)	Percent as Hb·NO	Percent as Hb-SNO
Venous	926[a]	97	3
Arterial	847[a]	63	37
Difference (Δ)	79*	−34	+34
		(venous loss)	(arterial gain)

Hb·NO, hemoglobin heme–nitric oxide complex.
[a] Approximately one molecule per 2500 molecules of hemoglobin.
*Not significant.

37% in arterial blood. This apparent arterial gain of approximately 34% in Hb-SNO is matched by an approximate 34% loss in Hb·NO, based on the venous content of Hb·NO. These data can be interpreted that ±0.3 µM of hemoglobin undergoes reciprocal shifts between Hb·NO on the venous side and Hb-SNO on the arterial side at each circulation.

Figure 6 ties these biological observations and the reactions of hemoglobin with NO into a schema for how shifts between Hb·NO and Hb-SNO content in blood might play a role in hemodynamics. Beginning on the venous side, the heme of the deoxyhemoglobin scavenges NO released into the microvasculature by endothelial cells, which determines the Hb·NO content of venous blood (eqn 1, Fig. 5). When this $HbFe^{2+}$·NO reaches the lung capillaries it is oxygenated, and methemoglobin is formed (by the reaction shown in eqn 2, Fig. 5). The methemoglobin ($HbFe^{3+}$) reacts with NO produced by lung capillary epithelial cells to form $HbFe^{2+}$-SNO (eqn 3, Fig. 5). When oxygenated, it then forms $HbFe^{2+}$·O_2-SNO, although some may also become oxidized to methemoglobin-SNO ($HbFe^{3+}$-SNO). These forms exit the lung in the arterial blood. On the basis of the hemodynamic effects reported in Table 3, the Hb-SNO species that exit the lung cause hypotension and thus increase perfusion, probably by vasodilation of capillaries. The mechanism for this reaction has not been determined, but one hypothesis is that in the microcirculation when Hb-Fe^{2+}·O_2-SNO gives up its oxygen to the tissues it is converted from an SNO carrier to an SNO donor, as indicated in Fig. 5 (bottom). Because this conversion must occur inside the erythrocyte where the major thiol is glutathione, the transnitrosation reaction probably favors formation of glutathione SNO (GS-SNO in Fig. 6). It is not known how GS-SNO leads to the relaxation of smooth muscle cells or pericytes when red blood cells containing $HbFe^{2+}$-SNO, GS-SNO, or both come into close contact with the capillary wall of vessels in the microcirculation. Although Fig. 6 is hypothetical, this schema could account for the shift in the proportions of Hb-SNO and Hb·NO, shown in Table 4, and for the hemodynamic effects resulting from injection of various hemoglobin derivatives, shown in Table 3.

The potential regulatory role of NO in hemodynamics is by no means the only example of such regulation. We have seen that NO complexed to heme and

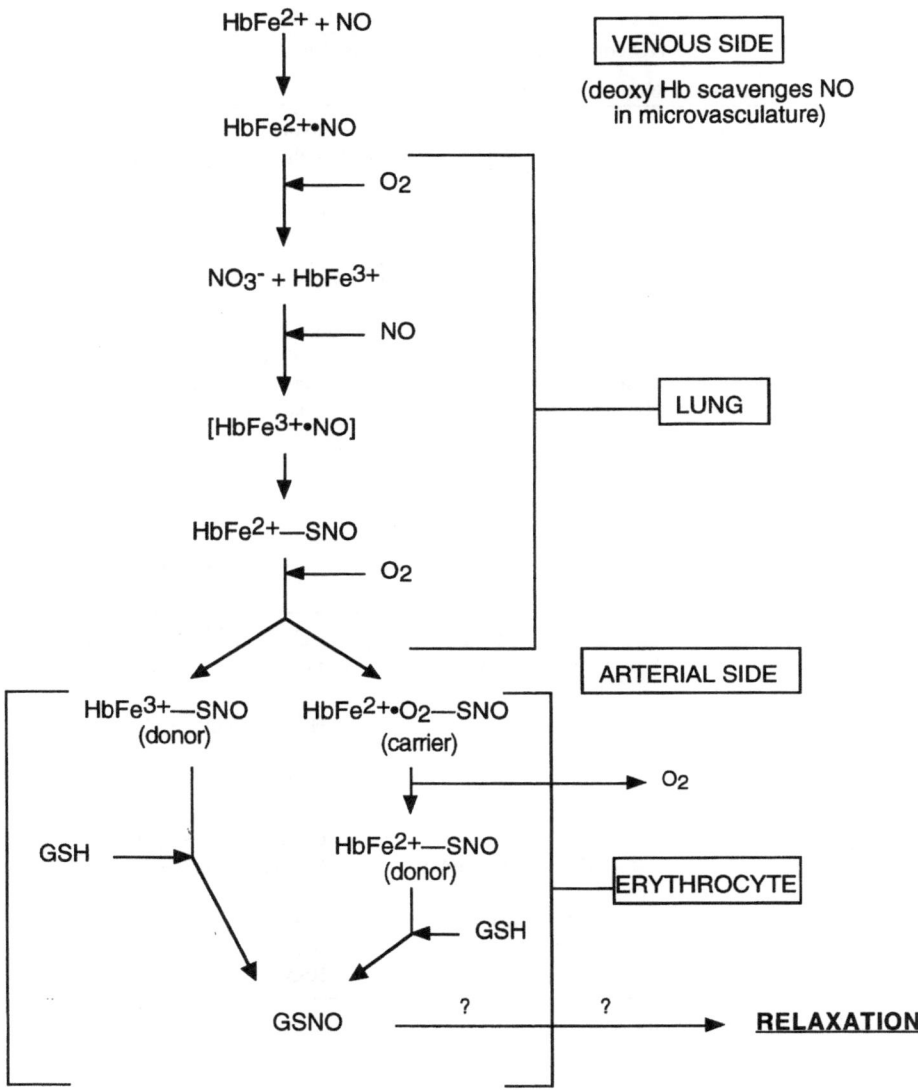

Fig. 6. Hypothetical scheme for shifts in the proportions of Hb•NO and Hb-SNO in the blood between the venous microcirculation (*venous side*) and the arterial microcirculation (*arterial side*) with each passage through the lung

TABLE 5. Examples of regulatory roles of NO

Fe·NO complexes		RSNO
Heme-Fe·NO	Fe-SNO	
G-cyclase	Aconitase	Glyceraldehyde-3-
COX-1, COX-2	Complex I	phosphate dehydrogenase
Hemoglobin	Complex II	PKC
	SoxR	γ-Glutamylcysteine-
		synthetase
		Hemoglobin
		OxyR

Fe·NO, nitric oxide complexed to protein-bound iron; RSNO, nitrosothiols; G-cyclase, guanylyl cyclase; COX, cyclooxygenase; Complex I, II, mitochondrial electron-transfer complexes; PKC, protein kinase C; SoxR, OxyR, bacterial transcription factors.

TABLE 6. Characteristics of *Escherichia coli* redox-sensitive transcription regulators

Characteristic	SoxR	OxyR
Size (kDa)	17 (acts as dimer)	35
Activators	O_2^-, NO	H_2O_2, RSNO
Active center	2[Fe—S] clusters	Cysteine 199
Activation reactions	$Fe^{2+} \rightarrow Fe^{3+}$	CysSH $\rightarrow$ CysS—OH
	Fe—SNO complex	CysSH $\rightarrow$ CysS—NO
Genes activated	SOD (Mn)	Catalase
		Alkyl hydroperoxidase
		RSNO metabolizing activity

SOD, superoxide dismutase.

nitrosation of a hemoglobin thiol most likely have a regulatory significance. Additional documented examples of regulation by NO complexed to bound Fe and by SNO formation of protein thiols are summarized in Table 5.

Two other examples, from bacteria, illustrate a potential regulatory role of NO that may be important for avoiding the "bad" side of NO, which is its well-documented potential for toxicity. They are the proteins termed SoxR and OxyR (Table 5). Both are transcription factors present in *Escherichia coli* that are redox sensitive. Their overall characteristics are summarized in Table 6. It is to be noted that each is activated by an active oxygen molecule (superoxide or hydrogen peroxide) and by NO and RSNO, respectively. The active centers of these proteins follow the theme of the previous discussion on hemoglobin, that is, a bound Fe^{2+} in SoxR to which NO can complex or easily oxidize (in this case the Fe is an iron–sulfur complex) and a specific thiol group in OxyR that is similarly readily oxidized to a sulfenic or sulfinic acid or transnitrosated (Hausladen et al., 1996). What is most interesting is that the gene expression controlled by these transcription factors involves important protective antioxidant proteins, such as superoxide dismutase, catalase, and peroxidase activities. These bacterial proteins provide a precedent for there being protective gene expression when NO or RSNO is present at toxic or deleterious levels ("nitrosative stress"). It is therefore worth considering that there are yet-to-be-discovered transcription factors in eukaryotic cells that are redox-sensitive to guard against the potential toxicity of NO and to ensure that the biological reactions of NO are confined to those that mediate its physiological functions.

References

Brenman JE, Chao DS, Xia H, Aldape K, Bredt DS (1995) Nitric oxide synthase complexed with dystrophin and absent from skeletal muscle sarcolemma in Duchenne muscular dystrophy. Cell 82:743–752

Geyer O, Podos SM, Mittag TW (1997) Nitric oxide synthase activity in tissues of the bovine eye. Graefes Arch Clin Exp Ophthalmol 235:786–794

Hausladen A, Privalle CT, Keng T, DeAngelo J, Stamler JS (1996) Nitrosative stress: activation of the transcription factor OxyR. Cell 86:719–729

Jia L, Bonaventura C, Stamler JS (1996) S-Nitrosohaemoglobin: a dynamic activity of blood involved in vascular control. Nature 380:221–226

Pillers DAM, Bulman DE, Weleber RG, Sigesmund DA, Musarella MA, Powell BR, Murphey WH, Westall C, Panton C, Becker LE, Worton RG, Ray PN (1993) Dystrophin expression in the human retina is required for normal function as defined by electroretinography. Nature Genet 4:82–86

Yamamoto R, Bredt DS, Snyder SH, Stone RA (1993) The localization of nitric oxide synthase in the rat eye and related cranial ganglia. Neuroscience 54:189–200

II. Aqueous Dynamics: Mechanisms of Ocular Hypertension and Glaucoma

Contractility of Trabecular Meshwork and Ciliary Muscle: Modulation by the NO/cGMP System

MICHAEL WIEDERHOLT, FRIEDERIKE STUMPFF, and NATALIE DÖRSCHNER

Introduction

Data indicate that nitric oxide (NO)/cyclic guanosine monophosphate (cGMP) can function as an intracellular signal transduction system. If specific cells contain both nitric oxide synthase (NOS) and guanylate cyclase activated by NO, intracellular cGMP can be regulated by substances that alter NOS activity and NO formation (Moncada et al. 1991; Moncada 1992; Nathan 1992).

It was possible to detect NOS-specific activity in ciliary processes and trabecular meshwork of the bovine eye (Geyer et al. 1993). NOS has now been described in the human outflow system and adjacent ciliary muscle (Nathanson and McKee 1995a). Furthermore, the reactivity of NOS decreased in patients with primary open-angle glaucoma. NADPH-diaphorase activity was subnormal in the trabecular meshwork and Schlemm's canal of glaucoma patients (Nathanson and McKee 1995b). In various animal models it has been shown that activation of the NO/cGMP system increases aqueous humor outflow and decreases intraocular pressure (Becker 1990; Busch et al. 1992; Schuman et al. 1994; Stein and Clack 1994; Behar-Cohen et al. 1996). Thus NO could be an important intracellular messenger in the regulation of aqueous humor dynamics via modulation of contractile properties of the trabecular meshwork.

It has been established that trabecular meshwork cells contain smooth muscle-specific actin and myosin filaments. Electrophysiological measurements demonstrated that cultured human and bovine trabecular meshwork cells express membrane voltages and excitability typical of smooth muscle cells (Coroneo et al. 1991; Lepple-Wienhues et al. 1994). Most importantly, the direct contractility of isolated meshwork strips from bovine eyes has been measured (Lepple-Wienhues et al. 1991a, 1991b, 1992; Wiederholt et al. 1993). A variety of functional receptors such as α- and β-adrenergic and endothelin receptors have been demonstrated in the trabecular meshwork (Lepple-Wienhues et al. 1991a, 1991b,

Institut für Klinische Physiologie, Universitätsklinikum Benjamin Franklin, Freie Universität Berlin, Hindenburgdamm 30, 12200 Berlin, Germany

1992; Wiederholt et al. 1993). The arguments for a direct contribution of the trabecular meshwork to regulating the outflow pathway in addition to the ciliary muscle has been reviewed elsewhere (Wiederholt et al. 1993; Wiederholt and Stumpff 1997).

Herein we summarize the effect of various nitro- and nonnitrate vasodilators on contractile properties of the isolated bovine trabecular meshwork (and ciliary muscle). The mechanisms of interaction of the NO/cGMP system on ion channels were examined using patch-clamp techniques.

Direct Measurement of Contractility of Trabecular Meshwork and Ciliary Muscle

The contractility of isolated ciliary muscle strips has been measured in several mammalian species including humans (for review: Wiederholt et al. 1993). In view of the smooth muscle-like properties of trabecular meshwork cells, we tried to obtain direct contractility measurements in isolated trabecular meshwork and ciliary muscle strips. To record isometric recordings at small forces (F), in the micronewton (μN) range, we developed a force–length–transducer system (Lepple-Wienhues et al. 1991a). Contractions of isolated strips are electrically counteracted via a lever-coil system in a magnetic field and an optoelectronic device. With this system, the contractile properties of trabecular meshwork could be measured directly for the first time (Lepple-Wienhues et al. 1991a). Our data with isolated strips indicate that cholinergic agonists (acetylcholine, carbachol, pilocarpine, aceclidine) may act on the outflow facility by both ciliary muscle contraction and a direct effect of the cholinomimetics on the trabecular meshwork itself (Lepple-Wienhues et al. 1991a; Wiederholt et al. 1996). A number of other substances were found to contract the trabecular meshwork as well (for review: Wiederholt and Stumpff 1997): high external K^+, α-adrenergic agonists, endothelin, thromboxane. Relaxant substances are β-adrenergic agonists, low external Ca^{2+}, Ca^{2+} channel blockers, prostaglandins of the E-type (EP-receptor agonists), ethacrynic acid. Thus a variety of drugs and substances (including most drugs used clinically to lower intraocular pressure) modulate the contractile properties of the trabecular meshwork. There are a number of similarities as well as fundamental differences between the contractility of the trabecular meshwork and ciliary muscle (for review: Wiederholt and Stumpff 1997).

Characterization of Channels by Patch-Clamp Techniques

Patch-clamp standard techniques were used as originally described by Neher and Sakman (Hamill et al. 1991). As cultured cells synthesize considerable amounts of extracellular matrix, the cell surface had to be enzymatically cleaned with trypsin before the experiments to obtain Giga-Ohm seals. Currents were recorded using an EPC 9 patch-clamp amplifier and 5- to 10-MΩ patch pipettes. For analysis of single-channels and whole-cell currents, data were analyzed by a special program described elsewhere (Berweck et al. 1994).

We were interested in studying channels in trabecular meshwork cells that are known to modify contraction/relaxation in smooth muscle cells. Among the various K$^+$ channels present in trabecular meshwork cells, we concentrated on the maxi-K$^+$ channel (K_{Ca}), which is an important regulator of the depolarization/hyperpolarization balance and thus contraction/relaxation in smooth muscle cells (Nelson and Quayle 1995). The whole-cell and single-channel figurations of the patch-clamp technique were used to study membrane conductance (Stumpff et al. 1996).

Effect of cGMP, Organic Nitrovasodilators and Nonnitrate Vasodilators on Contractility

It is now generally accepted that nitrovasodilators and NO activate intracellular soluble guanylate cyclase. The consequent increase in intracellular cGMP leads to protein phosphorylation and, in smooth muscle cells, relaxation. We tested various activation mechanisms of this system regarding the contractility of isolated bovine trabecular meshwork and ciliary muscle strips.

Figure 1, an original recording, shows that in our experimental setup we could demonstrate a relaxant effect of the membrane-permeable 8-bromo-cGMP at 10^{-4} mol/l in ciliary muscle and trabecular meshwork. At 10^{-5} mol/l, 8-bromo-cGMP had a relaxing effect only in trabecular meshwork. With in vitro experiments the concentration used is usually around 10^{-3} mol/l. Thus cGMP exhibited a strong potency to relax precontracted trabecular meshwork. The data are summarized in Table 1.

As the inhibitor to NOS we used L-arginine, which induced further contraction in tissues already maximally stimulated by carbachol (Table 1). The data indicate that in both trabecular meshwork and ciliary muscle there is a constant release of NO, the trabecular meshwork system being more sensitive.

The organic nitrates isosorbide dinitrate (ISDN) and isosorbide-5-mononitrate (ISMN) relaxed ciliary muscle and trabecular meshwork when tissues were precontracted by carbachol (Table 1). Whereas organic nitrates are known to activate soluble guanylate cyclase indirectly (via interaction with a thiol), nonnitrate vasodilators are known to be direct stimulators of the cyclase (Moncada et al. 1991; Moncada 1992; Nathan 1992). The nonnitrates sodium nitroprusside (SNP) and S-nitroso-N-acetyl penicillamine (SNAP) had the most effective relaxant effect on trabecular meshwork and ciliary muscle (Table 1). The powerful relaxing effect of the vasodilators demonstrates the presence of NOS in trabecular meshwork and ciliary muscle. In general, the relaxant effects were stronger in the former than in ciliary muscle.

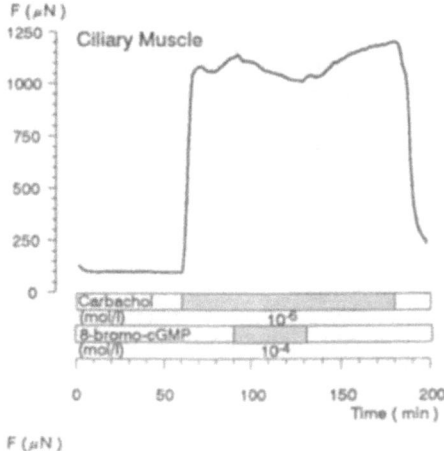

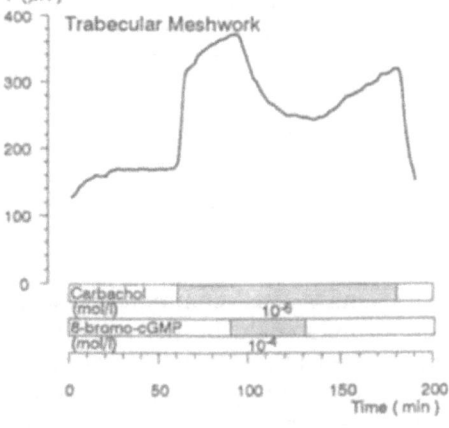

Fig. 1. Original recordings of isometric force development (F) in isolated strips of bovine ciliary muscle and trabecular meshwork. Carbachol induced a contraction that was partially relaxed by cGMP. (Data from Wiederholt et al. 1994)

Up to now, four isoforms of NOS have been characterized in numerous tissues. The enzyme may exist in soluble or particulate fraction of cells and may be constitutive or inducible (Moncada et al. 1991; Moncada 1992; Nathan 1992). To test for the presence of the constitutive isoform, a nitrovasodilator (ISDN) and a nonnitrate vasodilator (SNP) were tested in tissues that were neither stretched nor stimulated by carbachol. Because both substances significantly relaxed resting ciliary muscle and trabecular meshwork (data not shown; see Wiederholt et al. 1994) there is evidence of the constitutive and inducible isoforms in trabecular meshwork and ciliary muscle.

The constitutive isoform of NOS is regulated by intracellular calcium/calmodulin and other cofactors. In a second set of experiments we were interested in the mechanism of action of carbachol-induced contraction and smooth

TABLE 1. Effect of cGMP, organic nitrovasodilators, and nonnitrates on contractility

Substance (mol/l)	Relative contractility as percent of the force generated by carbachol	
	Trabecular meshwork	Ciliary muscle
Carbachol (10^{-6})	100	100
+ cGMP (10^{-5})	83.9 ± 4.4* (5)	98.2 ± 2.1 (7)
+ cGMP (10^{-4})	58.6 ± 5.4*** (7)	86.7 ± 1.4*** (8)
+ L-NAG (10^{-4})	118.5 ± 5.0* (6)	108.5 ± 1.0*** (7)
Carbachol (10^{-6})	100	100
+ ISDN (10^{-5})	83.9 ± 3.0** (7)	94.4 ± 0.8*** (7)
+ ISDN (10^{-4})	86.7 ± 3.7*** (9)	88.8 ± 5.7*** (10)
+ ISMN (10^{-4})	79.6 ± 3.6** (8)	92.2 ± 1.0*** (9)
Carbachol (10^{-6})	100	100
+ SNP (10^{-5})	65.9 ± 12.8** (6)	82.8 ± 1.2*** (8)
+ SNP (10^{-4})	38.6 ± 3.6*** (6)	55.5 ± 3.5*** (8)
+ SNAP (10^{-4})	37.1 ± 4.6*** (9)	68.4 ± 4.5*** (12)

Data are from Wiederholt et al. (1994).
Number of experiments are shown in parentheses.
cGMP, 8-bromo-cGMP; L-NAG, L-nitroarginine; ISDN, isosorbide dinitrate; ISMN, isosorbide-5-mononitrate; SNP, sodium nitroprusside; SNAP, S-nitro-N-acetyl penicillamine.
*$P < 0.05$; **$P < 0.01$; ***$P < 0.001$ (versus 100% contraction induced by carbachol).

muscle relaxation elicited by NO-generating agents. A final common pathway for modulation of contractility by carbachol and NO could be the activity of intracellular calcium. In cultured human and bovine trabecular meshwork (and ciliary muscle), muscarinic agonists such as carbachol induce depolarization of the membrane voltage, action potentials, an increase in intracellular calcium, and contractions (Lepple-Wienhues et al. 1991a, 1991b, 1992, 1994; Wiederholt et al. 1994, 1996).

In smooth muscle cells carbachol opens nonselective cation channels that are permeable to Na^+, K^+, and Ca^{2+} (Benham et al. 1985). A similar mechanism of action seems to be present in trabecular meshwork and ciliary muscle (Lepple-Wienhues et al. 1991b, 1992), as the most specific blocker of nonselective cation channels currently known, flufenamic acid (Gögelein et al. 1990; Isenberg 1993), significantly relaxed these two tissues precontracted by carbachol. Approximately 50% of the carbachol-induced contraction could be relaxed by flufenamic acid, the relaxation being more pronounced in trabecular meshwork. In addition to the effects of carbachol on nonselective cation channels, muscarinic agonists possibly modify various other channels and transporters. A direct effect of NO on ion channels without requiring cGMP has been postulated (Bolotina et al. 1994). To test for a possible interaction between muscarinic agonists and activation of NO/cGMP on nonselective cation channels, the following experiments were performed.

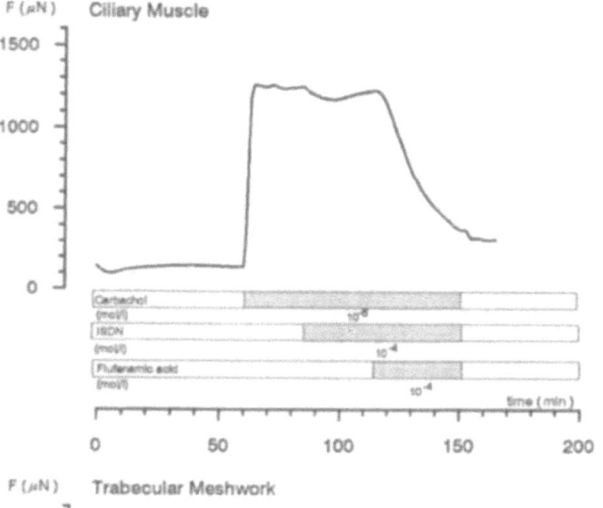

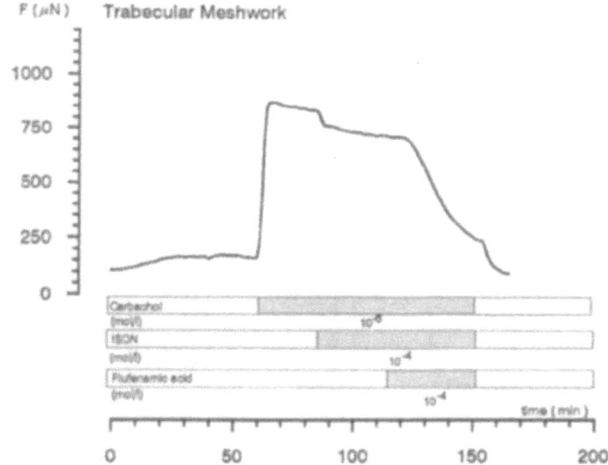

FIG. 2. Original recordings of contractility when isosorbide dinitrate (*ISDN*) was applied to relax the precontracted tissues partially. Application of the specific blocker of nonselective cation channels, flufenamic acid, further reduced the contraction and thus induced pronounced relaxation. (Data from Wiederholt et al. 1997)

Figures 2 and 3 are original recordings obtained when ISDN was given first followed by the nonselective cation channel blocker flufenamic acid (Fig. 2) or, vice versa, the channel blocker was given first and then ISDN (Fig. 3). As can be seen in the original recordings and in the summary of the data (Figs. 4, 5) the effect of ISDN on relaxation in precontracted tissue was independent of the blockade of nonselective cation channels by flufenamic acid. Also the effect of flufenamic acid was independent of ISDN. The absolute amount of relaxation induced by the substances are in the range of the data reported before (Wiederholt and Stumpff 1997; Wiederholt et al. 1997). It is concluded that there is no interaction between the NO/cGMP system and the carbachol-modulated nonselective cation channel pathway in terms of inducing relaxation of trabecular meshwork and ciliary muscle tissue.

FIG. 3. Original recordings of tissues precontracted by carbachol. The blocker of nonselective cation channels, flufenamic acid, induced relaxation. In both tissues isosorbide dinitrate was an additional relaxant

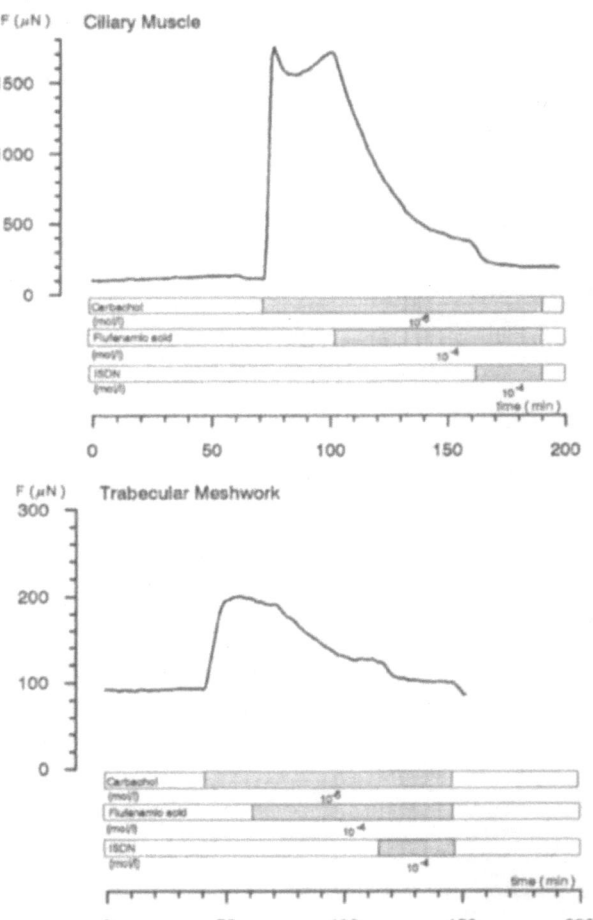

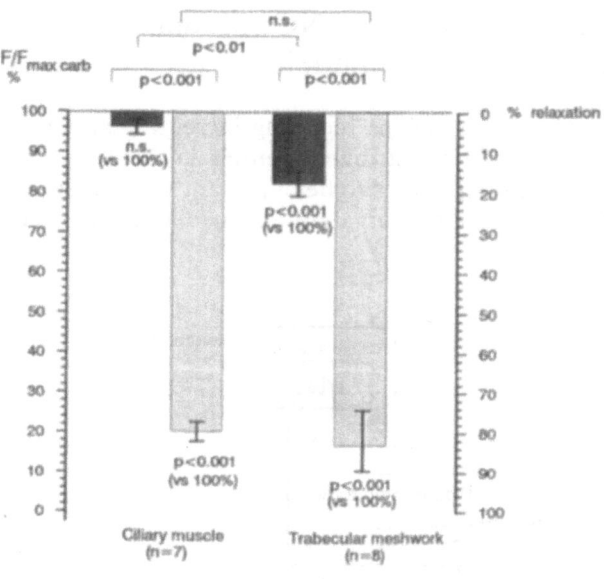

FIG. 4. Summary of the interaction of isosorbide dinitrate and flufenamic acid on tissues precontracted by carbachol. Isosorbide dinitrate was given first, followed by flufenamic acid

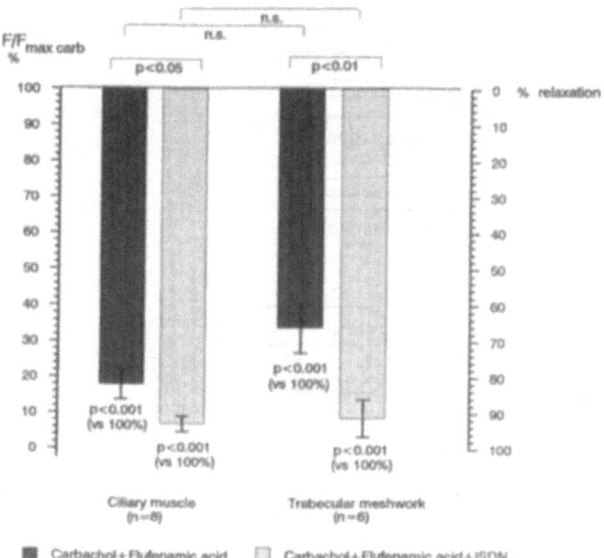

FIG. 5. Summary of the interaction of isosorbide dinitrate and flufenamic acid on tissues precontracted by carbachol. Flufenamic acid was given first, followed by isosorbide dinitrate

Relaxation of Trabecular Meshwork by NO: Mediation Via Calcium-Dependent Maxi-K⁺ Channels

The maxi-K⁺ channel in smooth muscle cells is receiving increasing attention (Nelson and Quayle 1995). This channel is distributed densely in the cell membrane of various smooth muscle cells and seems to be important for the regulation of smooth muscle tone. The channel is the target protein mediating the effect

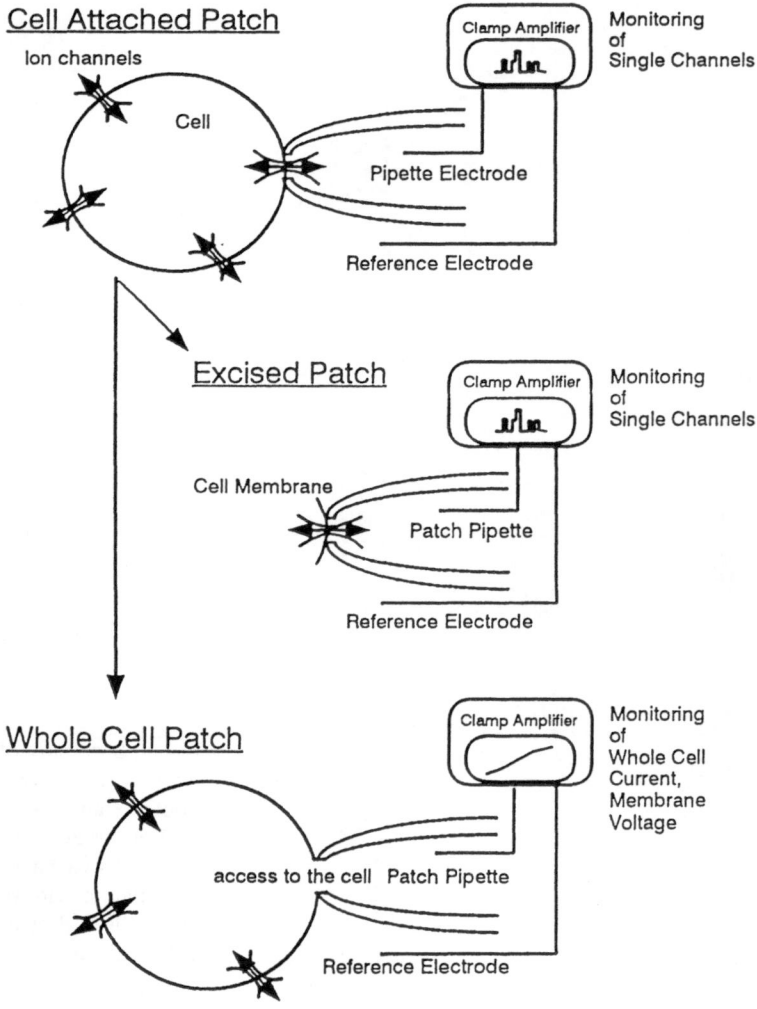

FIG. 6. Various patch–clamp techniques

TABLE 2. Characterization of a Ca^{2+}-dependent maxi-K^+ channel in trabecular meshwork cells

Single channel configuration
  Inside-out patches
    High conductance for K^+: $326 \pm 4\,pS$ (n = 10)
    Negligible conductance for Na^+: $0.9 \pm 1\,pS$ (n = 10)
    Open probability (NP_o) voltage-dependent
    Open probability increased at intracellular calcium $>10^{-6}\,mol/l$
    Adenosine triphosphate (ATP) increased open probability
  Outside-out patches
    Charybdotoxin ($10^{-8}\,mol/l$) reversibly reduced open probability
    Tetraethyl ammonium (TEA) reduced open probability

Cell-attached recordings
    Channel opened at physiological voltage levels
    Channel conductance was not modified by chloride

Whole-cell recordings
    Strong outward current, voltage-dependent
    Charybdotoxin led to a reduction of outward current by $35 \pm 5\%$
    Current induced by increase of intracellular calcium (by ionomycin) could be partly blocked by charybdotoxin
    TEA dose-dependently blocked outward current
    Application of the lipid-diffusible cGMP analog 8-bromo-cGMP ($10^{-3}\,mol/l$) increased outward current by $290 \pm 75\%$
    Charybdotoxin ($10^{-7}\,mol/l$) led to a reduction of the cGMP-induced current by $44 \pm 7\%$

Data are from Stumpff et al. (1997).

of various relaxant substances. We concentrated on the function of the maxi-K^+ channel in trabecular meshwork cells (Stumpff et al. 1996, 1997).

The various patch-clamp techniques used are represented in Fig. 6. Excised inside-out and outside-out patches were used to characterize the maxi-K^+ channel on a single channel level. The overall characteristics of the cell membrane were described by cell-attached and whole-cell recordings. The characteristics of this Ca^{2+}-dependent maxi-K^+ channel are summarized in Table 2. The channel exhibited voltage-dependent activation and inactivation and could be blocked by extracellular charybdotoxin, tetraethyl ammonium (TEA), and internal Ba^{2+}. Charybdotoxin has been described as a specific blocker of maxi-K^+ channels. Increased intracellular Ca^{2+} activated the channel. These characteristics are typical of maxi-K^+ channels observed in various smooth muscle cells (Nelson and Quayle 1995). It is of importance that intracellular Ca^{2+} elevated to the physiological range and depolarization increased the open probability (NP_o) of this channel. Activation of the maxi-K^+ channel leads to an increase of K^+ efflux, which may counteract depolarization of the membrane voltage. Substances that block the maxi-K^+ channel should depolarize the membrane voltage and, via a secondary increase in intracellular calcium via voltage-dependent Ca^{2+} channels, increase the contractility of smooth muscle cells.

Relaxation induced by cGMP, nitrovasodilators, and nonnitrates was more pronounced in trabecular meshwork than in ciliary muscle. Clarification of the mechanism of action of the NO/cGMP system on trabecular meshwork seems to be of considerable interest. The importance of the NO/cGMP system has been documented in transformed human trabecular meshwork (contractile) cells, in which sodium nitroprusside increased intracellular cGMP (Pang et al. 1994). In patch-clamp experiments performed on cells derived from the same preparation of bovine trabecular meshwork used in the contractility measurements mentioned above, we were able to demonstrate a high density of maxi-K^+ channels. Application of the membrane-permeable 8-bromo-cGMP to the external bathing solution led to a large increase in outward current (Fig. 7, Table 2). The specific blocker of the maxi-K^+ channels, charybdotoxin, led to a significant reduction of the current induced by cGMP. Thus elevation of cytosolic cGMP stimulates Ca^{2+}-dependent K^+ channels in trabecular meshwork cells, implying an involvement of this channel in relaxation of the trabecular meshwork. Figure

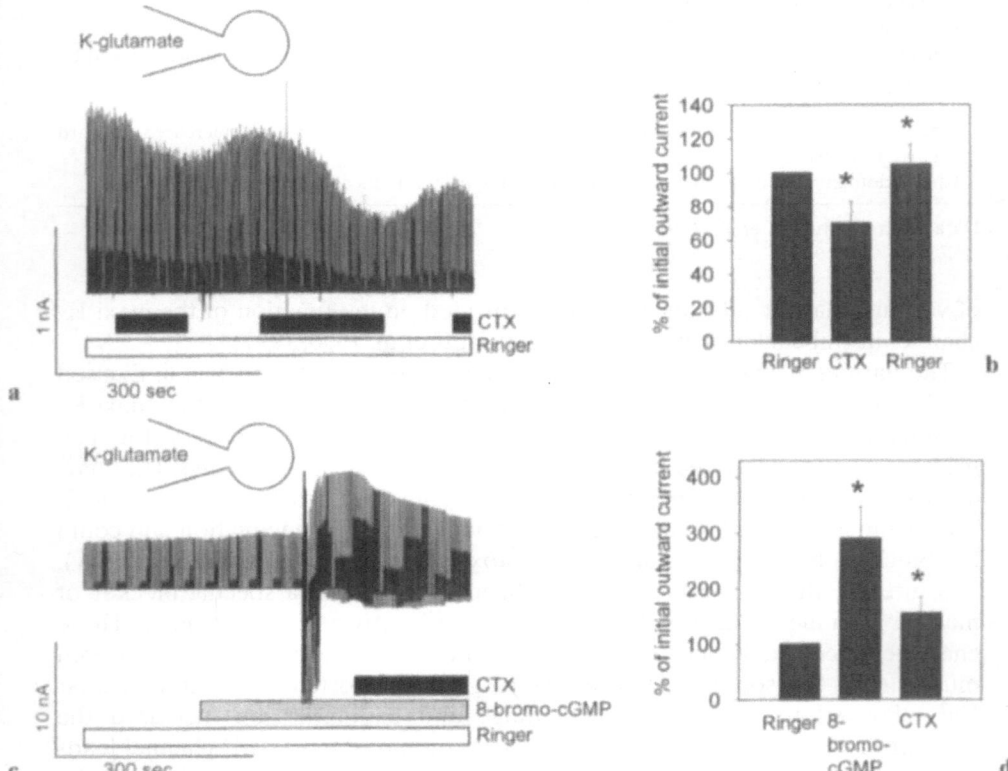

FIG. 7. **a** Charybdotoxin (*CTX*) reduces outward current in the whole-cell configuration. **b** Summary of data from five experiments such as those in **a**. **c** Application of 8-bromo-cGMP induces outward current, which is blockable by charybdotoxin. **d** Summary of data from four experiments such as those in **c**

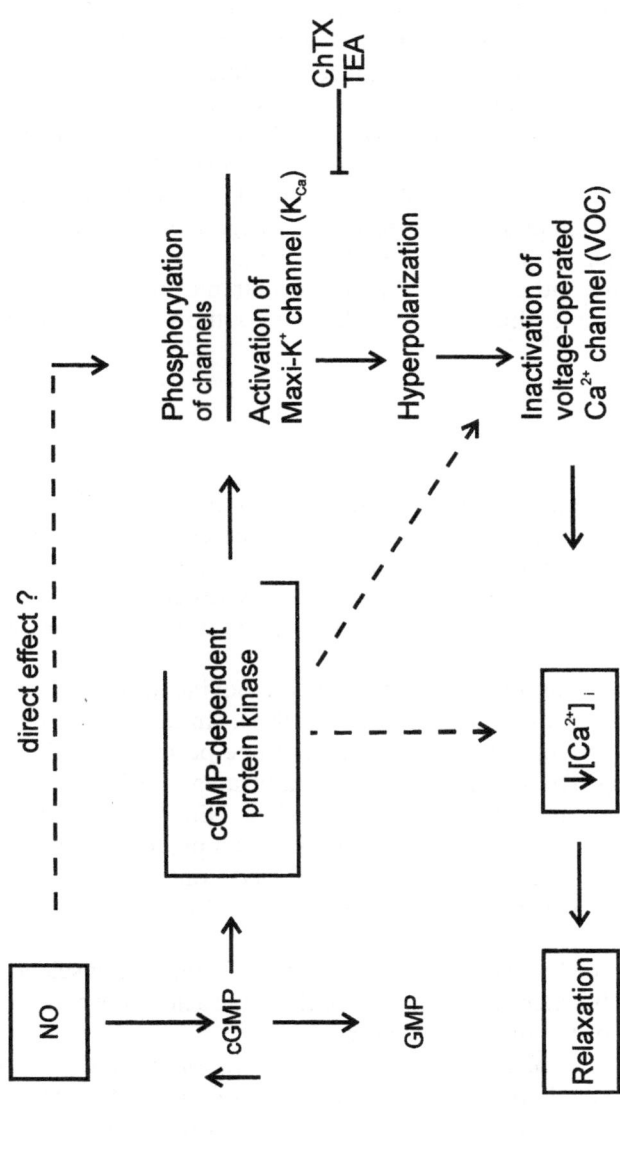

Fig. 8. Proposed mechanism of action of NO/cGMP-induced relaxation by modulation of maxi-K+ channels. This scheme was modified according to Archer et al. (1994) and Bolotina et al. (1994). *cGMP*, cyclic guanosine monophosphate; *ChTx*, charybdotoxin; *TEA*, tetraethyl ammonium

8 summarizes the proposed mechanism of action of nitric oxide on relaxation of the trabecular meshwork. Data indicate that cGMP-dependent protein kinases are able to activate the maxi-K^+ channel in smooth muscle cells (Archer et al. 1994; Nelson and Quayle 1995).

In addition to this cGMP-dependent mechanism a direct effect of NO on the maxi-K^+ channel has been postulated (Bolotina et al. 1994). This channel is known as a target for various substances modulating smooth muscle contractility. After activation of this channel potassium leaves the cell through this channel, inducing hyperpolarization, which in turn leads to inactivation of voltage-operated Ca^{2+} channels and finally decreases intracellular calcium, inducing relaxation of the trabecular meshwork. In trabecular meshwork cells, the maxi-K^+ channel seems to serve as a negative-feedback mechanism, regulating membrane voltage/intracellular calcium and hence the balance between contraction and relaxation. In trabecular meshwork, NO may modulate contractility and thus aqueous humor outflow by modulating this channel.

Conclusion

We presented evidence elsewhere for contractile properties of human and bovine trabecular meshwork cells. In the bovine eye, at least the trabecular meshwork per se is directly involved in the regulation of aqueous humor outflow (Wiederholt et al. 1995). The present study tested the functional significance of the guanylate cyclase/cGMP system on trabecular meshwork and ciliary muscle.

Isometric tension was measured directly on bovine trabecular meshwork (and ciliary muscle) strips (Wiederholt et al. 1994). An increase in intracellular cGMP induced by direct application of membrane-permeable cGMP, organic nitrovasodilators (ISDN, 5-ISMN), and nonnitrates (SNP, SNAP) evoked relaxation in strips precontracted by carbachol. Nitrovasodilators also had significant relaxing activity on strips without induced precontraction. Inhibition of NO synthase increased the carbachol-induced contraction. A continuous release of NO under resting and precontracted conditions seems to maintain a relaxing effect.

Contractility induced by carbachol is partially mediated by nonselective cation channels. A typical blocker of this channel, flufenamic acid, did not interfere with the relaxing effect of the nitrovasodilators.

The whole-cell and single-channel configurations of the patch-clamp technique were used to study membrane conductance of cultured trabecular meshwork cells. In whole-cell experiments a charybdotoxin-sensitive outward current was observed that increased in amplitude when intracellular calcium was increased. In single-channel experiments trabecular meshwork cells were shown to have a channel with a $326 \pm 4\,pS$ conductance for potassium and negligible conductance for sodium. The activity of the channel was dependent on membrane voltage and intracellular calcium. In the whole-cell experiments application of membrane-permeable cGMP increased the outward current. The cGMP-dependent current could be totally blocked by charybdotoxin.

Trabecular meshwork cells express maxi-K^+ channels in high density, comparable to that in other smooth muscle cells. These channels are involved in regulating the balance between relaxation and contraction. Relaxation induced by an increase in cGMP is mainly mediated via the maxi-K^+ channels. Elevation of internal cGMP levels leads to an activation of maxi-K^+ channels and thus to relaxation.

Acknowledgment. This work was supported by Deutsche Forschungsgemeinschaft (grant Wi 328/19).

References

Archer SL, Huang JMC, Hampl V, Nelson DP, Shultz PJ, Weir EK (1994) Nitric oxide and cGMP cause vasorelaxation by activation of a charybdotoxin-sensitive K channel by cGMP-dependent protein kinase. Proc Natl Acad Sci USA 91:7583–7587

Becker B (1990) Topical 8-bromo-cyclic GMP lowers intraocular pressure in rabbits. Invest Ophthalmol Vis Sci 31:1647–1649

Behar-Cohen FF, Goureau O, D'Hermies F, Courtois Y (1996) Decreased intraocular pressure induced by nitric oxide donors is correlated to nitrite production in the rabbit eye. Invest Ophthalmol Vis Sci 37:1711–1715

Benham CD, Bolten TB, Lang RJ (1985) Acetylcholine activates an inward current in single mammalian smooth muscle cells. Nature 316:345–347

Berweck S, Lepple-Wienhues A, Stöss M, Wiederholt M (1994) Large conductance calcium-activated potassium channels in cultured retinal pericytes under normal and high-glucose conditions. Pflugers Arch 427:9–16

Bolotina VM, Najibi S, Palacino JJ, Pagano PJ, Cohen RA (1994) Nitric oxide directly activates calcium-dependent potassium channels in vascular smooth muscle. Nature 368:850–853

Busch MJWM, van Oosterhout JGM, Hoyng PFJ (1992) Effects of cyclic nucleotide analogs on intraocular pressure and trauma-induced inflammation in the rabbit eye. Curr Eye Res 1:5–13

Coroneo MT, Korbmacher C, Stiemer B, Flügel C, Lütjen-Drecoll E, Wiederholt M (1991) Electrical and morphological evidence for heterogeneous populations of cultured bovine trabecular meshwork cells. Exp Eye Res 52:375–388

Geyer O, Podos SM, Mittag TW (1993) Nitric oxide synthase: distribution and biochemical properties of the enzyme in the bovine eye. Invest Ophthalmol Vis Sci 34:826

Gögelein H, Dahlem D, Englert HC, Lang HJ (1990) Flufenamic acid, mefenamic acid and niflumic acid inhibit single nonselective cation channels in the rat exocrine pancreas. FEBS Lett 268:79–82

Hamill OP, Marty A, Neher E, Sakman B, Sigworth FJ (1991) Improved patch-clamp techniques for high-resolution current recording from cells and cell-free membrane patches. Pflugers Arch 1981:85–100

Isenberg G (1993) Nonselective cation channels in cardiac and smooth muscle cells. In: Siemen D, Hescheler J (eds) Nonselective cation channels: pharmacology, physiology and biophysics. Birkhäuser Verlag, Basel, pp 247–260

Lepple-Wienhues A, Stahl F, Wiederholt M (1991a) Differential smooth muscle-like contractile properties of trabecular meshwork and ciliary muscle. Exp Eye Res 53:33–38

Lepple-Wienhues A, Stahl F, Willner U, Schäfer R, Wiederholt M (1991b) Endothelin-evoked contractions in bovine ciliary muscle and trabecular meshwork: interactions with calcium, nifedipine and nickel. Curr Eye Res 10:983–989

Lepple-Wienhues A, Stahl F, Wunderling D, Wiederholt M (1992) Effects of endothelin and calcium channel blockers on membrane voltage and intracellular calcium in cultured bovine trabecular meshwork. German J Ophthalmol 1:159–163

Lepple-Wienhues A, Rauch R, Clark AF, Grässmann A, Berweck S, Wiederholt M (1994) Electrophysiological properties of cultured human trabecular meshwork cells. Exp Eye Res 59:305–311

Moncada S (1992) The L-arginine:nitric oxide pathway. Acta Physiol Scand 145:201–227

Moncada S, Palmer RMJ, Higgs EA (1991) Nitric oxide: physiology, pathophysiology, and pharmacology. Pharmacol Rev 43:109–142

Nathan C (1992) Nitric oxide as a secretory product of mammalian cells. FASEB J 6:3051–3064

Nathanson JA, McKee M (1995a) Identification of an extensive system of nitric oxide-producing cells in the ciliary muscle and outflow pathway of the human eye. Invest Ophthalmol Vis Sci 36:1765–1773

Nathanson JA, McKee M (1995b) Alterations of ocular nitric oxide synthase in human glaucoma. Invest Ophthalmol Vis Sci 36:1774–1784

Nelson MT, Quayle JM (1995) Physiological roles and properties of potassium channels in arterial smooth muscle. Am J Physiol 268:C799–C822

Pang IH, Shade DL, Clark AF, Steely HT, DeSantis L (1994) Preliminary characterization of a transformed cell strain derived from human trabecular meshwork. Curr Eye Res 13:51–63

Schuman JS, Erickson K, Nathanson JA (1994) Nitrovasodilator effects on intraocular pressure and outflow facility in monkeys. Exp Eye Res 58:99–105

Stein PJ, Clack JW (1994) Topical application of a cyclic GMP analog lowers IOP in normal and ocular hypertensive rabbits. Invest Ophthalmol Vis Sci 35:2765–2768

Stumpff F, Strauss O, Wagner U, Wiederholt M (1996) Cultured bovine trabecular meshwork cells possess maxi-K-channels in high density. Invest Ophthalmol Vis Sci 37:S205

Stumpff F, Strauss O, Boxberger M, Wiederholt M (1997) Characterization of maxi-K-channels in trabecular meshwork and their activation by cGMP. Invest Ophthalmol Vis Sci 38:1883–1892

Wiederholt M, Stumpff F (1997) The trabecular meshwork and aqueous humor reabsorption. In: Civan MM (ed) Current topics in membranes, Vol 45. The eye's aqueous humor: from secretion to glaucoma. Academic, San Diego, pp 163–202

Wiederholt M, Lepple-Wienhues A, Stahl F (1993) Contractile properties of trabecular meshwork and ciliary muscle. In: Lütjen-Drecoll E (ed) Basic aspects of glaucoma research III. Schattauer, Stuttgart, pp 287–306

Wiederholt M, Sturm A, Lepple-Wienhues A (1994) Relaxation of trabecular meshwork and ciliary muscle by release of nitric oxide. Invest Ophthalmol Vis Sci 35:2515–2520

Wiederholt M, Bielka S, Schweig F, Lütjen-Drecoll E, Lepple-Wienhues A (1995) Regulation of outflow rate and resistance in the perfused anterior segment of the bovine eye. Exp Eye Res 61:223–234

Wiederholt M, Schäfer R, Wagner U, Lepple-Wienhues A (1996) Contractile response of the isolated trabecular meshwork and ciliary muscle to cholinergic and adrenergic agents. German J Ophthalmol 5:146–153

Wiederholt M, Dörschner N, Groth J (1997) Effect of diuretics, channel modulators, and signal interceptors on contractility of the trabecular meshwork. Ophthalmologica 211:153–160

Origin and Function of Nitrergic Nerves in the Human Eye: Morphological Aspects

ERNST R. TAMM and ELKE LÜTJEN-DRECOLL

Introduction

Signals transmitted by the ocular part of the autonomic nervous system regulate important auxiliary systems that are necessary to maintain the basic function of the eye, the perception of light. A considerable number of physiological and morphological studies during the past few years have provided strong evidence that numerous ocular autonomic nerves contain the neuronal isoform of nitric oxide synthase (NOS), the enzyme that synthesizes nitric oxide (NO), and likely use NO as a neurotransmitter. In the eye, such nitrergic nerves are involved in regulating the blood supply to the various intraocular tissues and appear to take part in accommodation and the circulation of aqueous humor.

Studies on the tissue-specific localization of nitrergic nerves have been made possible by the synthesis of specific antibodies against neuronal NOS (Bredt et al. 1992; Klatt et al. 1992). Another method to identify NOS in nerve cells and their axons uses the NADPH-diaphorase (NADPH-D) reaction, an NADPH-dependent reduction of nitroblue tetrazolium to the dark-blue, water-insoluble dye nitroblue tetrazolium formazan (Dawson et al. 1991; Hope et al. 1991). Although the NADPH-D reaction is sensitive for detecting NOS, it appears to be not entirely specific. There are other NADPH-dehydrogenating enzymes that can reduce nitroblue tetrazolium (Blottner et al. 1995), and NOS represents only a fraction of the total cellular NADPH-D activity in nonfixed tissues (Tracey et al. 1993). The specificity of the NADPH-D reaction for identifying neuronal NOS can be increased by formaldehyde fixation, which inactivates some of the NADPH-D activity not related to NOS (Matsumoto et al. 1993; Beesley 1995; Blottner et al. 1995). Under these conditions, NADPH-D is a convenient method for screening neuronal NOS, although any positive results must be confirmed with more specific methods. Our knowledge about the localization and origin of

Department of Anatomy, University of Erlangen-Nürnberg, Universitätsstr. 19, 91054 Erlangen, Germany

nitrergic nerves in the human eye, which is the subject of this review, has been obtained by a combination of NOS immunohistochemistry, NADPD-D-histochemistry, and electron microscopy.

Nitrergic Vascular Nerves in the Posterior Eye Segment

The posterior segment of the human eye is supplied by two vascular systems: the retinal and choroidal vasculatures. The retinal vessels are branches of the central retinal artery and nourish the inner two-thirds of the retina, whereas the choroidal vessels derive from 10–20 short posterior ciliary arteries that penetrate the sclera at the posterior pole. Their capillary bed, the choriocapillaris, supplies the avascular fovea and the outer parts of the retina including the photoreceptors. The two vascular systems differ markedly in terms of their fine structure. The endothelial cells of the retinal capillaries are sealed with tight junctions and form the blood–retina barrier, whereas the choriocapillaris is fenestrated and highly permeable to low-molecular-weight substances.

The first studies using antibodies against NOS to investigate the tissue-specific localization of NOS-immunoreactive (IR) nerves in general visualized a dense network of such axons in the choroid of the rat eye (Dawson et al. 1991; Bredt et al. 1992). More specific studies on the distribution of NOS-IR nerves in the rat eye confirmed these findings (Yamamoto et al. 1993). In addition, they showed that NOS-IR axons are present in many parts of the rat eye, with the choroid as their predominant localization. In this species the choroidal nitrergic axons emerge at the optic nerve head and form thick bundles that are associated with the posterior ciliary arteries (Flügel et al. 1994) (Fig. 1a). They run in parallel with the ciliary arteries and follow each of their arborizations. Continuously, smaller axons leave the paravascular nerve bundles and join a delicate, dense perivascular network surrounding the vascular wall of all choroidal arteries and arterioles. Only toward the periphery of the choroid does the density of the nitrergic perivascular network decrease. Yamamoto et al. (1993) described an extension of the perivascular nitrergic nervous network into the rat choriocapillaris. Such a dominant nitrergic innervation of the choroid seems to be a distinct characteristic feature of the eyes of most mammalian species, as elaborate nitrergic nervous networks, albeit sometimes less regular and dense than in the rat eye, have been demonstrated in the rabbit, cat, tree shrew, owl monkey, cynomolgus monkey, and human (Bergua et al. 1993; Flügel et al. 1994; Flügel-Koch et al. 1994). Moreover, reports on the presence of a similar dense nitrergic innervation in the choroid of birds (Fitzgerald and Reiner 1993; Bergua et al. 1995, 1996) have indicated that a nitrergic innervation of the choroid might be common to the vertebrate eye in general.

In the human eye a striking difference distinguishes its nitrergic choroidal innervation from that of most other mammalian species: the presence of a widespread plexus of NOS- and NADPH-D-positive intrinsic nerve cells (Bergua et al. 1993; Flügel et al. 1994) (Fig. 1b,c). Most of these cells have a polygonal

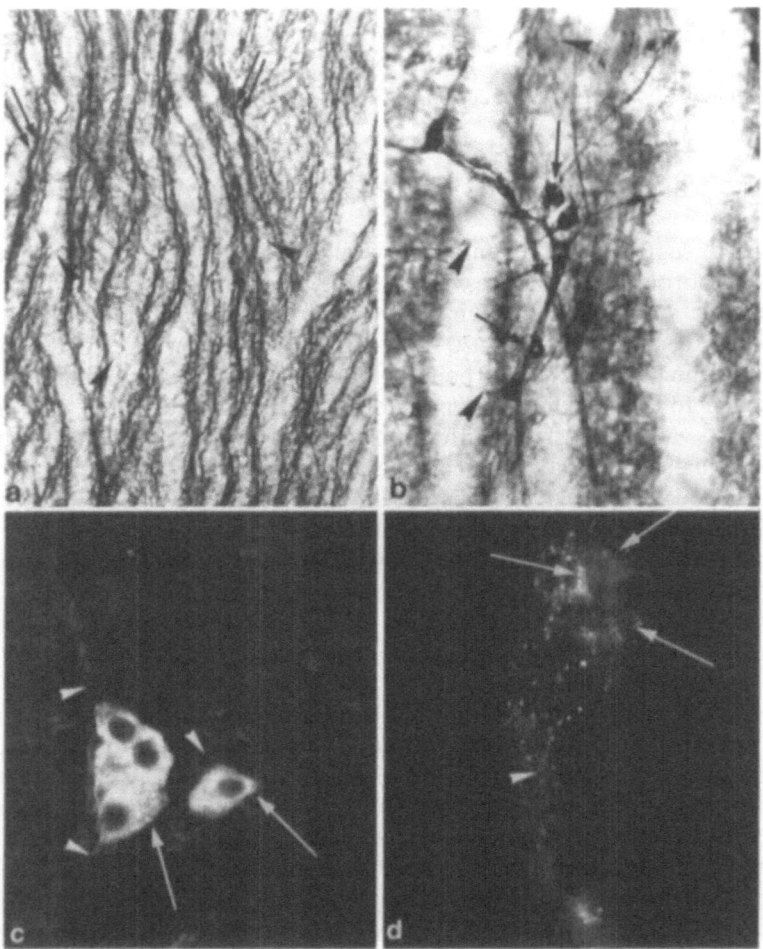

FIG. 1. Nitrergic innervation of the choroid. **a** Whole mount of a rat choroid stained for NADPH-diaphorase (NADPH-D). The temporal region in the vicinity of the optic nerve head is shown. Bundles of positively stained nerve fibers follow the course of the arterial vessels (*double arrows*). Small axons originate from these bundles, forming a delicate network around the vascular wall (*arrowheads*). (×80) **b** Whole mount of a human choroid (temporal region) stained for NADPH-D. Positively stained ganglion cells are shown (*arrows*). These cells are connected with each other by stained axons. Additional nerve fibers lead to the perivascular fiber network (*arrowheads*). (×51) **c** Frozen tangential section of human choroid (temporal quadrant) stained immunocytochemically with antibodies against nitric oxide synthase (NOS). Strong positive labeling is seen in the cytoplasm of the ganglion cells (*arrows*). Weak staining for NOS is also found for the axons of these nerve cells (*arrowheads*). (×220) **d** Frozen tangential section through the temporal quadrant of a human choroid after incubation for synaptophysin. Two ganglion cells are seen surrounded by positively labeled varicosities (*arrows*). Positive staining is also present in small vesicles lying in the cytoplasm of axons (*arrowhead*). (×270) (From Flügel et al., 1994, with permission of Lippincott-Raven)

perikaryon, with diameters ranging from 10 to 40μm. The cells appear either solitary or, more often, clustered in groups of 2–10 cells. Most of the ganglion cells are located adjacent to the wall of larger choroidal arteries in the outer layers of the choroid near the sclera; they are never observed in the layer of the choriocapillaris. Individual cells appear to be mostly multipolar. Groups of ganglion cells are connected to each other by NOS/NADPH-D-positive axons. Although these connecting axons show no apparent association with the choroidal vasculature, thin axons that originate from them join the network of perivascular nerve fibers around arteries and arterioles. Some of the axons continue to the choriocapillaries where they appear to be in contact with pericytes.

In whole-mount preparations of the choroid from five individual human eyes, 1555–2579 NOS/NADPH-D-positive ganglion cells were counted per eye (Flügel et al. 1994). The highest number of these cells (approximately 1000) and the largest cells (20–40μm) were found in the central temporal area of the choroid (below the fovea). Toward the periphery the number of ganglion cells decreased, as did the size of the cells. Most of the peripheral ganglion cells have diameters less than 10μm. In general, the diameter of choroidal ganglion cells seems to increase with age (Kurus 1955; Flügel et al. 1994), which might be due to an accumulation of lipofucsin granules, a typical event in the aging process of neurons. Choroidal ganglion cells are innervated: Immunohistochemistry with antibodies against synaptophysin, a transmembranous glycoprotein specifically localized to the membranes of synaptic vesicles (Wiedenmann and Franke 1985), and electron microscopy have visualized numerous varicose terminals closely associated with the perikarya of the cells (Fig. 1d). It is not clear whether these terminals derive only from the axons of other choroidal ganglion cells or from sources outside the eye. Such an extrinsic innervation might come from one or more of the cranial autonomic ganglia known to project to the eye (the parasympathetic ciliary and pterygopalatine ganglia, the sympathetic superior cervical ganglion, and the sensory trigeminal ganglion) (Stone et al. 1987). In this case, choroidal nerve cells would represent a third neuron in a rather complex nerve pathway. Another possibility is that choroidal neurons are postganglionic to neurons in the brain stem (e.g., in classical parasympathetic centers such as the nucleus of Edinger-Westphal or the superior salivatory nucleus).

Some of the choroidal ganglion cells in the human eye express vasoactive intestinal polypeptide (VIP)-IR (Miller et al. 1983; Flügel et al. 1994). It is not known whether VIP-IR and NOS-IR are co-localized in choroidal neurons as they are in other autonomous ganglion cells outside the eye (Kummer et al. 1992; Talmage and Mawe 1993; Ding et al. 1995).

As with other nervous structures, human choroidal ganglion cells can be visualized with the supravital fluorescent dye 4-Di-2-ASP (Bergua et al. 1994). This elaborate system of intrinsic nerve cells in the human choroid, which in many structural details resembles the intrinsic ganglia of the intestinal nervous system, was discovered more than 100 years ago by Heinrich Müller (1859b) and was described in detail somewhat later by Iwanoff (1874). Although choroidal ganglion cells were subject to some histologic studies in more recent decades

(Lauber 1936b; Kurus 1955; Feeney and Hogan 1961; Castro-Correira 1967), their function remained unclear, and some authors of textbooks dealing with the anatomy of the human choroid do not even mention them.

Not all NOS-positive axons that innervate choroidal vessels in the human eye are derived from intrinsic choroidal ganglion cells. Because NOS-positive axons are already found in the wall of the short posterior ciliary arteries before and shortly after they penetrate the sclera, a significant number of such axons appear to project to the human choroid from extrinsic extraocular NOS-containing ganglion cells.

In addition to perivascular nerves and choroidal ganglion cells, NADPH-D activity has been demonstrated in the vascular endothelial cells of almost all vessels in the human choroid (Bergua 1995). In contrast to neuronal staining, NADPH-D activity in endothelial cells is not homogeneous but appears as punctate blue staining in the cytoplasm. In our material, NADPH-D activity in choroidal vascular endothelial cells of human eyes was seen in arteries and arterioles, albeit with considerable weaker intensity than in the perivascular nerves. No or only extremely weak NADPH-D activity was seen in the choriocapillaries. In addition, in the eyes of the same human donors, retinal vascular endothelial cells invariably showed much stronger NADPH-D activity than choroidal vascular endothelial cells. Although these results should be confirmed with more specific methods (e.g., immunohistochemistry with specific antibodies), it seems reasonable to assume that the endothelial NADPH-D staining reflects the presence of the endothelial isoform of NOS and that this enzyme is more active in retinal than in choroidal vascular endothelial cells.

Nitric oxide released by perivascular nerves relaxes vascular smooth muscle and has been shown to be responsible for nonadrenergic-noncholinergic (NANC) innervation in many parts of the circulation (Sanders and Ward 1992; Schmidt and Walter 1994; Morris et al. 1995; Toda 1995). Choroidal blood flow is modulated by vasodilative NANC nerves, demonstrated following intracranial stimulation of the facial nerve in rabbits, cats, and monkeys (Stjernschantz and Bill 1980; Nilsson et al. 1985). In contrast, no change in choroidal blood flow was observed after stimulation of the oculomotor nerve (Stjernschantz and Bill 1979). Studies have provided evidence that choroidal nitrergic nerves are at least partly responsible for NANC vasodilation. In pigs, electrical stimulation of the isolated long posterior ciliary artery, which supplies the choroid, retina, and anterior uvea in this species, caused NANC vasodilation that could be blocked with inhibitors of NOS (Su et al. 1994). Similar results were reported for posterior ciliary arteries of humans (Nyborg and Nielsen 1994). Using microspheres in anesthetized dogs, Deussen et al. (1993) showed a 40% decrease in choroidal blood flow following intravenous infusion of NOS inhibitors. Laser flowmetry in cats demonstrated that infusion with NOS inhibitors caused an decrease in choroidal blood flow and attenuated an acetylcholine-induced increase (Mann et al. 1995). In rabbits, the increase in choroidal blood flow following stimulation of the facial nerve was shown to be reduced or abolished by NOS inhibition at low stimulation frequencies (Nilsson 1996). In rabbits, as in other species including humans, the axons of the facial nerve synapse on nerve cells in the ptery-gopalatine ganglion that project their axons to the eye (Ruskell 1965, 1970a, 1970b, 1971). Indeed, a large number of pterygopalatine ganglion cells in rat, dog, pig, and monkey stain for NOS, NADPH-D, or both (Morris et al. 1993; Suzuki et al. 1993; Yamamoto et al. 1993; Yoshida et al. 1993, 1994; Sienkiewicz et al. 1995). Pterygopalatine ganglionectomy in rats significantly reduces the number of peripheral NOS-IR nerve fibers in the eye (Yamamoto et al. 1993). In humans the pterygopalatine ganglion may be the main source of those choroidal ni-trergic axons that do not derive from the intrinsic choroidal ganglion cells but project from extraocular nerve cells to the choroid. We showed that approxi-mately 70% of human pterygopalatine ganglion cells stain for NOS and NADPH-D, whereas in the human ciliary ganglion fewer than 1% of the neurons are NOS/NADPH-positive (Tamm and Lütjen-Drecoll 1996a). Similar findings were reported for the monkey ciliary ganglion (Sun et al. 1994). Another source of ocular nitrergic axons in humans appears to be the trigeminal ganglion; in the ophthalmic portion of this ganglion 17%–19% of the ganglion cells are NOS- and NADPH-D-positive (Tamm and Lütjen-Drecoll 1996a).

Neurotransmitters other than NO may also be involved in NANC vasodila-tion of the choroid. In rabbits, although NOS inhibition reduced choroidal vasodilation following facial nerve stimulation only at low stimulation frequen-cies, it was not affecting it at high frequencies (Nilsson 1996) suggesting effects of neurotransmitters other than NO. Likely candidates are VIP and the struc-turally related pituitary adenylate cyclase-activating polypeptide (PACAP). Both peptides are found around choroidal vessels (in numerous neurons of the ptery-gopalatine ganglion) and are potent vasodilators of choroidal vessels, at least in some mammalian species (Nilsson and Bill 1984; Stone et al. 1987; Nilsson 1994; Mulder et al. 1995; Wang et al. 1995).

Electrical stimulation of the nucleus of Edinger-Westphal in pigeons caused an increase in choroidal blood flow, an effect that was significantly reduced fol-lowing administration of NOS inhibitors (Zagvazdin et al. 1996). The latter study provided additional evidence for the vasodilative effect of NO released from choroidal nerves, but it also showed that the pathway of such nerves in birds is different from that in mammalian eyes. The nucleus of Edinger-Westphal in pigeons provides preganglionic input to "choroidal" ganglion cells in the ciliary ganglion. These choroidal ganglion cells send their axons to the vessels in the avian choroid; however, their structure is different from those that supply the ciliary muscle (Dryer and Chiapinelli 1985; De Stefano et al. 1993), and they are not found in the mammalian ciliary ganglion (Ehinger et al. 1983; May and Warren 1993; Zhang et al. 1994b).

In summary, the choroid of many species seems to be under the influence of a basal release of NO, which maintains a vasodilatory tone of choroidal vessels; and it seems reasonable to assume that nitrergic choroidal nerve terminals con-tribute a substantial amount of it. In humans these terminals derive from axons of intraocular intrinsic ganglion cells in the choroid and, in addition, probably from extraocular ganglion cells in the pterygopalatine and trigeminal ganglia.

There is no doubt that the choroidal vasculature is of crucial importance to the delivery of nutrients to the retina. Experiments in anesthetized cats and monkeys show that the amount of oxygen supplied by the choroid and consumed in the retina is about 80% and 65%, respectively (Alm and Bill 1972, 1973). In pigs about 60% of the oxygen and 75% of glucose are delivered to the retina by the choroidal vessels (Törnquist and Alm 1979). In cats the oxygen tension in those parts of the retina (including the photoreceptors) that are supplied by the choroid is more than twice as high as in parts supplied by the retinal vessels (Alder et al. 1983). In primates the avascular fovea is totally dependent on a supply from the choroid. Still, the physiological role of the dense nitrergic vasodilative innervation of choroidal vessels is less clear, as blood flow through the choroid is so high that it exceeds the metabolic requirements. The oxygen extraction from choroidal vessels is low; the arteriovenous difference in oxygen concentration measures only 2%–3% (Alm and Bill 1970). In addition, the fenestrated choriocapillaris is highly permeable to glucose or larger macromolecules (Bill 1968; Bill et al. 1980). Studies on the choroidal vasculature in cats and pigs indicate that the net extraction of oxygen and glucose is maintained at a normal level, even after large alterations in choroidal blood flow (Alm and Bill 1970; Törnquist and Alm 1979). An intriguing possibility is that the physiological role of the high rate of choroidal blood flow is not so much nutrition but prevention of a damaging increase in intraocular temperature (Parver et al. 1980; Bill 1985; Alm 1992). Such an increase might occur while observing bright objects emitting light with high intensity, especially in primates where the light is focused on the fovea. Under these conditions the dense nitrergic vasodilative innervation might be important for a reflexive increase in choroidal blood flow (Parver et al. 1982, 1983). In support of this hypothesis, López-Costa et al. (1995) reported on an increase in NADPH-D staining in choroidal nerves of rats exposed to continuous illumination for 1–10 days, an effect that was reversible after 20 days in darkness. Although these data are intriguing, confirmation with other methods seems necessary. Despite their reliability for qualitative observations, histochemical stains such as the NADPH-D reaction are difficult to quantitate as they are subject to a complex number of parameters that are hard to control.

An open question remains: Why do humans have a dense network of choroidal ganglion cells, and other mammalian species such as the rat and rabbit do not (Flügel et al. 1994)? A possibility is that the presence of choroidal ganglion cells is related to the development of a centralized area of vision, the fovea centralis, which among mammals is present only in humans and higher primates (Rohen 1962, 1982). Indeed, almost half of the ganglion cells in the human choroid are located in the central temporal region of the choroid, adjacent to the fovea (Flügel et al. 1994). This distribution corresponds exactly to that area in rhesus monkeys (*Macaca mulatta*, a species with fovea centralis) where choroidal blood was shown to be highest (Alm and Bill 1973; Alm 1992).

To clarify these issues we studied the presence of ganglion cells in a considerable number of mammalian species (Flügel-Koch et al. 1994). Only in cynomolgus monkeys (*Macaca fascicularis*, a species with a well-developed fovea, was a plexus of NOS/NADPH-D-positive choroidal ganglion cells found comparable to that in humans. In contrast, all the other species including the cat, pig, tree shrew, and owl monkey (*Aotes trivirgatus*, a nocturnal monkey without a fovea) showed a dense nitrergic innervation of choroidal vessels but no or only a few choroidal ganglion cells.

This association between ganglion cells and the fovea centralis seems not to be restricted to mammals, as a distinct population of choroidal ganglion cells has also been described in the choroid of birds (Fitzgerald and Reiner 1993; Bergua et al. 1995, 1996). Anatomically, the most conspicuous and complex foveas are found in birds, some species of which have two foveas in each eye (Franz 1934). The functional significance of an association between choroidal ganglion cells and fovea centralis remains to be established. The fovea centralis is supplied only by choroidal vessels, and the presence of a resident autonomous nerve cell plexus might be more advantageous in mediating vasodilative reflexes when light is focused on the fovea. Another hypothesis is based on the fact that a fovea is present only in eyes of species with a well-developed accommodation system (Rohen 1982; Flügel-Koch et al. 1994). In these eyes a more sophisticated control of vasodilative reflexes mediated by a choroidal ganglion cell plexus might be favorable, as changes in blood volume should change choroidal thickness, thereby influencing the position of the fovea centralis and hence visual acuity. In human eyes the position of the fovea might also be modulated by a distinct network of choroidal nonvascular smooth muscle/myofibroblast-like cells (Flügel-Koch et al. 1996). These cells, which are most abundant in the choroid underneath the fovea, are in contact with autonomous nerve endings of a yet unidentified nature.

In marked contrast to their abundance around choroidal vessels, autonomic perivascular NOS/NADPH-D-positive nerves are not present in the human retina (Chakravarthy et al. 1994), which is in accordance with findings in the rat and rabbit (Osborne et al. 1993; Yamamoto et al. 1993; Chakravarthy et al. 1994). The absence of NOS-positive autonomic nerves around vessels in the retina is not surprising, as in mammals (with the exception of the rabbit) retinal vessels beyond the surface of the optic disc are in general not supplied by any kind of autonomic vascular nerves (Laties 1967; Ye et al. 1990). However, perivascular NADPH-D-positive nerve fibers have been demonstrated around the extraocular central retinal artery of the dog, monkey, and human (Toda et al. 1994, 1996; Bergua 1996). In dogs the source of these nerves appears to be the pterygopalatine ganglion (Toda et al. 1993). In vitro experiments have provided evidence that NO released from nitrergic nerves relaxes isolated strips of monkey and dog central retinal artery (Toda et al. 1994, 1996). In rabbits, stimulation of the facial nerve significantly increased retinal blood flow, an effect that could be abolished by NOS inhibition (Nilsson 1996). In the dog intraarterial infusion of nicotine caused dilation of arterioles in the fundus region, which could be abolished by hexamethonium and NOS inhibition, suggesting the involvement of nitrergic axons (Toda et al. 1994). Furthermore, nicotine does not dilate retinal arteries in dogs where the pterygopalatine ganglion was damaged by ethanol injection, indicating the pterygopalatine ganglion as the source of these axons (Toda et al. 1993). In summary, then, nitrergic nerves may influence retinal blood vessels by an effect on extraocular vessels.

In addition, some experiments indicate that NO is released intraocularly in the retina from sources other than the autonomic nerves. Microinjection of an NOS inhibitor in the retina of miniature pigs induces arteriolar vasoconstriction by 45% (Donati·et al. 1995). Retinal pericytes, which typically surround retinal capillaries (Frank et al. 1990), relax in vitro in response to sodium nitroprusside, an NO donor (Haefliger et al. 1994). The endothelial cells of most of the retinal vessels in humans, rats, and rabbits stain for NOS and NADPH-D (Yamamoto et al. 1993; Chakravarthy et al. 1994; Perez et al. 1995). Reverse transcription-polymerase chain reaction (RT-PCR) experiments have demonstrated the expression of mRNA for endothelial NOS in bovine retinal microvascular endothelial cells (Chakravarthy et al. 1994). In addition, some NOS-IR retinal nerve cells (mostly amacrine cells) in the human and rat have been reported to associate closely with the basal lamina covering endothelial cells and pericytes of retinal capillaries (Roufail et al. 1995). Electron microscopic immunocyto-chemical studies of the subcellular localization of NOS have demonstrated that NO synthesis takes place in both the perikaryon of nerve cells and the peripheral terminal axon (Llewellyn-Smith et al. 1992). If NO is synthesized in the perikaryon of retinal nerve cells, it might easily diffuse to neighboring vascular cells. Thus a local mechanism coupled to intrinsic retinal activity would contribute to regulation of the retinal circulation. Such a functional coupling of amacrine-derived nerve processes and retinal vessels has already been suggested for the action of classical neurotransmitters and neuropeptides (Ye et al. 1990). Support for this hypothesis comes from comparison with the brain, where neuronal processes associated with intraparenchymal capillaries are thought to regulate blood flow (Lou et al. 1987). Interestingly, flickering light increases blood flow in the cat and monkey retina and in the cat optic nerve head (Riva et al. 1991; Wang et al. 1995) and the formation of NO in the pig retina (Donati et al. 1995). NOS inhibition blocks flicker-induced vasodilatation in the cat but not in the monkey (Wang et al. 1995).

The NOS-IR intrinsic retinal nerve cells, which seem to be predominantly amacrine cells, are numerous and have been studied in several species (Yamamoto et al. 1993; Perez et al. 1995). A detailed discussion of their nature, spatial distribution, and functional significance is beyond the scope of this review.

Nitrergic Vascular Nerves in the Anterior Eye Segment

The anterior ciliary arteries are the major source of blood supply to the anterior segment of the primate eye (Wilcox et al. 1980). They penetrate the sclera in the region of the ciliary muscle to form an arterial circle in the muscle and, together with the two long posterior ciliary arteries, form the major arterial circle of the iris, the main blood supply to the iris and ciliary processes (Morrison and Van

Buskirk 1983; Funk and Rohen 1990). In humans the vascular smooth muscle cells in the media of both arterial circles are in contact with a network of perivascular NADPH-D-positive nerve fibers (Tamm et al. 1995a). We found that this network appears to be denser in areas where arterioles to the ciliary body emerge from the major arterial circle (Fig. 2a).

A regulatory function of these arterioles has been hypothesized based on scanning and transmission electron microscopic studies, which showed segmental luminal constrictions (Morrison and Van Buskirk 1984; Funk and Rohen 1988) and a two-layered vascular wall of myocytes in association with numerous nerve endings (Funk 1991). A characteristic nitrergic innervation is not present around other vessels in the iris or ciliary body. We observed solitary NADPH-D-positive axons in the stroma of the iris, but they were not in intimate contact with iridial vessels (Fig. 2c). In contrast, iridial NADPH-D-positive axons appear to continue to the iris sphincter muscle, where they are found in close contact with some of the muscle cells (Fig. 2d). In the ciliary muscle, some NADPH-D positive axons are occasionally observed in contact with smaller arterioles. Similarly, some scattered NADPH-positive axons are seen near the capillaries of the ciliary processes (Fig. 2b). Overall, the nitrergic vascular innervation is far less pronounced in the anterior uvea of the human eye than in the posterior uvea. Comparable findings were reported for the rat eye (Yamamoto et al. 1993). Most of the nitrergic fibers around the arterial circles probably are derived from sources outside the eye, as they are already observed around those segments of the anterior ciliary arteries that penetrate the sclera. Only in the ciliary muscle do axonal processes of neighboring ciliary muscle nerve cells (discussed later) occasionally contribute to the perivascular network of the two arterial circles. Similar to some of the NOS/NADPH-D-positive axons in the posterior uvea, nitrergic perivascular nerves in the anterior uvea most probably are derived from the pterygopalatine and trigeminal ganglia. A vasodilative role for perivascular nitrergic nerves in the anterior uvea is likely. Neuronal NO was shown to be involved in NANC vasodilatation of the isolated intraocular part of the bovine ciliary artery that supplies the ciliary body (Wiencke et al. 1994). In vivo experiments in monkeys, cats, and rabbits showed a NANC vasodilation in the iris and ciliary body following intracranial stimulation of the facial nerve (Stjernschantz and Bill 1980; Nilsson et al. 1985; Nilsson 1996). In rabbits this vasodilation following facial nerve stimulation was significantly reduced following NOS inhibition (Nilsson 1996).

In human eyes the endothelial cells of almost all vessels in the anterior uvea stain for NADPH-D, which likely reflects the presence of endothelial NOS (Bergua 1995). This endothelial stain is pronounced in the intramuscular capillaries of the ciliary muscle. Endothelial NOS is released as endothelial-derived relaxing factor following stimulation of vascular endothelial cells with acetylcholine (Palmer et al. 1987). Ciliary muscle cells in human and monkey eyes express dense cholinergic innervation (Kaufman 1992; Tamm and Lütjen-Drecoll 1996c). During contraction, acetylcholine released from the numerous nerve endings in the muscle may spread not only to ciliary muscle cells but also to the

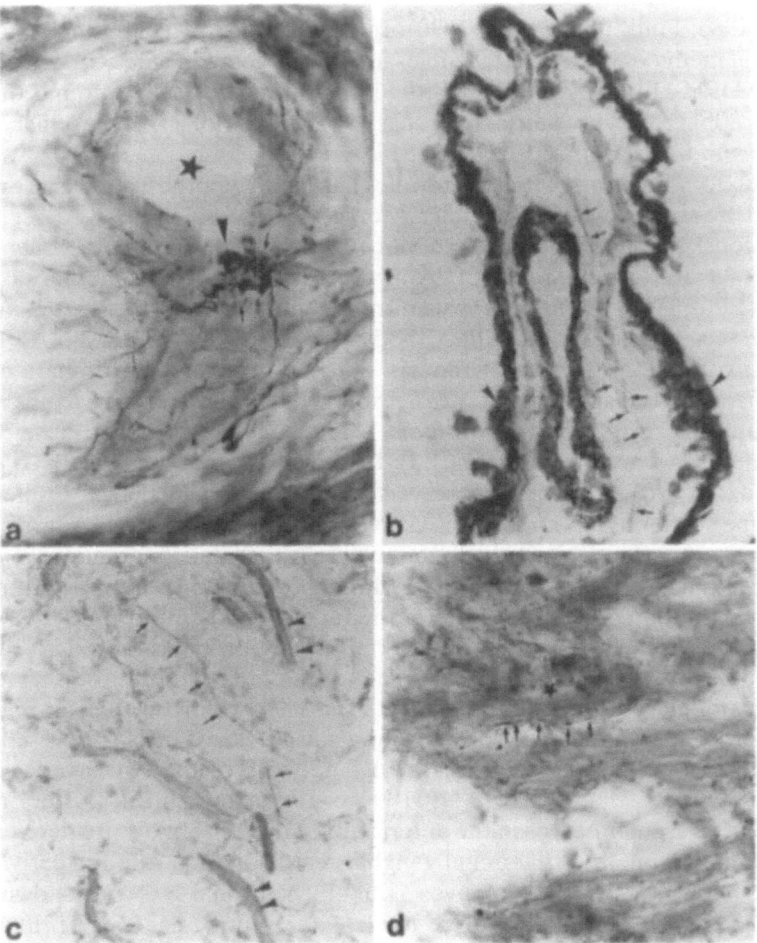

FIG. 2. Nitrergic axons in human ciliary body and iris. **a** The network of perivascular NADPH-D-positive axons in the wall of the major arterial circle of the iris (*star*) is most dense (*arrows*) in areas where arterioles to the ciliary body emerge (*arrowhead*). (×300) **b** Axons in the stroma of the ciliary processes stain for NADPH-D (*arrows*). In addition, the nonpigmented ciliary epithelium shows a positive NADPH-D reaction (*arrowheads*). (×250) **c** Tangential section through the stroma of the iris. Solitary NADPH-D-positive axons (*arrows*) are observed in the stroma, but they do not form a perivascular NADPH-D-positive nervous network around iris vessels (*arrowheads*). (×250) **d** Tangential section through the iris sphincter (*star*). Several smooth muscle cells in the iris sphincter are in close contact with a varicose NADPH-D-positive axon (*arrows*). (×250)

endothelial cells, causing dilation of the intramuscular capillaries. Such a local mechanism could help to maintain a sufficient supply of oxygen and nutrients during contraction. Ciliary muscle cells contain considerably more mitochondria than other smooth muscle cells (Tamm and Lütjen-Drecoll 1996c), and contraction of the ciliary muscle in response to acetylcholine or carbachol is highly sensitive to a decrease in oxygen supply (van Alphen et al. 1962; Törnqvist 1967b).

Nitrergic Nerves in the Ciliary Muscle

In the human eye uveal ganglion cells are not confined to the choroid but are present in the ciliary muscle as well. Similar to the choroidal ganglion cells, the nerve cells in the ciliary muscle were first discovered more than 100 years ago by Müller (1859a). They were subsequently mentioned in several histological studies of human ciliary muscle (Krause 1861; Iwanoff 1874; Lauber 1936a; Krümmel 1938; Kurus 1955; Bryson et al. 1966; Castro-Correira 1967), although some authors investigating innervation of the human ciliary muscle did not observe them or even questioned their existence (Boeke 1933; Hirano 1941; van der Zypen 1967). A possible explanation is the fact that most of the nerve cells in the ciliary muscle are considerably smaller than other mammalian peripheral nerve cells (e.g., those of the enteric nervous system) (Stöhr 1957; Furness et al. 1988). About 70% of ciliary muscle nerve cells have a longitudinal diameter of only 10–14 µm, whereas 30% measure 20–30 µm (Tamm et al. 1995a) (Fig. 3). Both soma and dendrites of the nerve cells stain for the neuronal marker PGP 9.5 (Wilkinson et al. 1989) and for NOS and NADPH-D, indicating that they are nitrergic in nature (Tamm et al. 1995a) (see Figs. 5d, 6, 7A,B).

In contrast to choroidal nerve cells, ciliary muscle nerve cells usually appear singly; only occasionally are two ganglion cells found close to each other. Ganglion cells in the human ciliary muscle are characteristically localized between the muscle bundles of the reticular and circular portion, areas that are most important for accommodation (Fig. 3). They are not found within individual muscle bundles or in the anterior longitudinal portion of the muscle.

Electron microscopy shows that ciliary muscle ganglion cells are surrounded by flat processes of glial cells, which also surround numerous preterminal unmyelinated axons and several terminal boutons (Fig. 4). The boutons form axosomatic contacts with the perikaryon and contain mitochondria and numerous small agranular (40–60 nm) and some large granular vesicles (60–120 nm) (Fig. 5). Ganglion cells may also express spine-like or lamellar dendritic processes where axodendritic synapses are formed. Large ganglion cells are multipolar and express numerous lamellar dendrites or several long and short filamentous processes (Fig. 6A,B,C). Small ganglion cells have a smoother surface, some short filamentous processes, and one or two long filamentous processes (Fig. 6D). In favorable sections, presumably axonal processes of both cell types

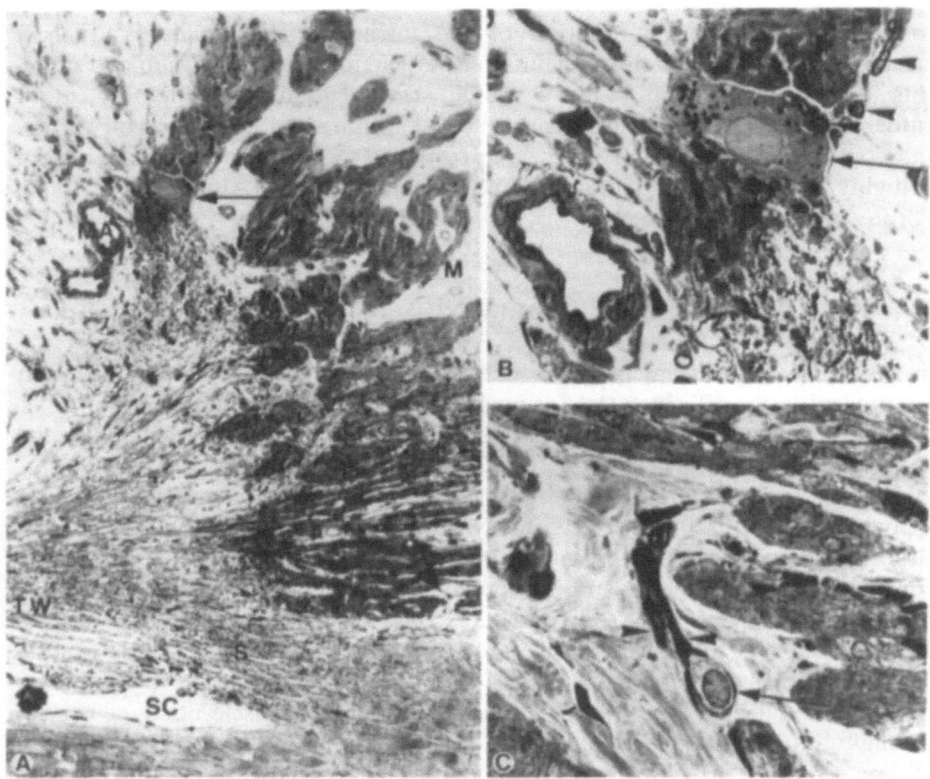

FIG. 3. Nerve cells in the human ciliary muscle. **A** Oval neuron (*arrow*) with a longitudinal diameter of 30 μm is situated between the muscle bundles of the circular portion. *M*, ciliary muscle; *S*, scleral spur; *SC*, Schlemm's canal; *TW*, trabecular meshwork; *MA*, major arterial circle of the iris (donor age 56 years). (×330) **B** Higher magnification of **A** shows the nerve cell (*arrow*) to be characterized by a large euchromatic nucleus, lipofuscin particles, and clear cytoplasm. Myelinated axons (*arrowheads*) are seen in close proximity to the nerve cell. **C** Small (longitudinal diameter 11 μm) ciliary muscle neuron (*arrow*) between the muscle bundles of the reticular portion. Similar to large neurons, small neurons are commonly associated with myelinated axons (*arrowheads*) (donor age 71). (×1000) (Semithin sections, Richardson's stain) (From Tamm et al. 1995a, with permission of Lippincott-Raven)

can be traced for 80–100 μm (Fig. 6E). The axons of several ganglion cells located close to the major arterial circle of the iris seem to contribute to the NOS/NADPH-D-positive perivascular network of this vessel and may be involved in NANC vasodilatation. Most of the ciliary muscle ganglion cells, however, have axons that run close and parallel to the ciliary muscle bundles and exhibit periodic swellings, suggesting varicosities, along the muscle cells.

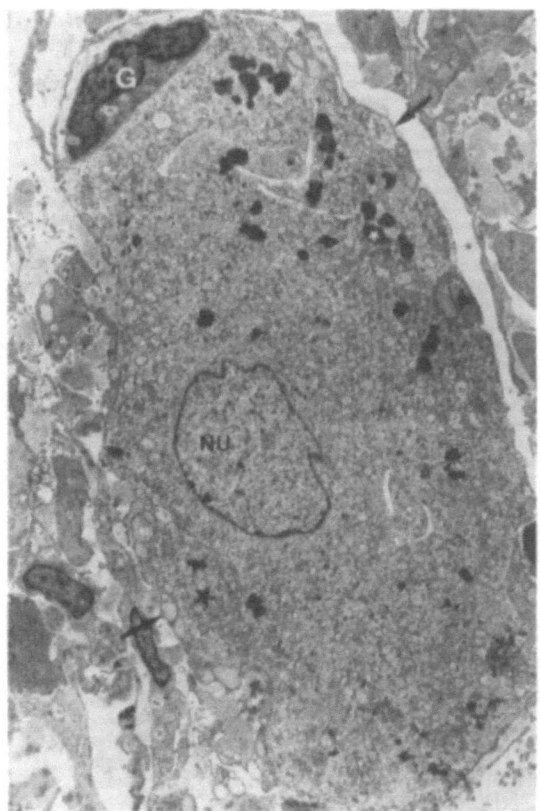

FIG. 4. Electron micrograph of a nerve cell in the ciliary muscle (same neuron as in Fig. 3a,b). The perikaryon of the neuron contains numerous mitochondria and profiles of short rough endoplasmic reticulum (rER) cisterns and free polysomes (*black star*). Scattered throughout the cytoplasm are large granular vesicles and irregularly shaped lipofuscin granules (*white star*). The neuron is surrounded by flat processes of glia cells (*G*), which also surround numerous preterminal, unmyelinated axons (*arrows*). *NU*, nucleus. (×7600) (From Tamm et al., 1995a, with permission of Lippincott-Raven)

Occasionally, neighboring ganglion cells are connected to each other by axonal processes (Fig. 6F).

In a quantitative analysis of three human eyes, 17–32 NOS/NADPH-D-positive nerve cells were counted per millimeter of circumferential ciliary muscle width. Given a total circumferential ciliary muscle width of about 36mm (Tripathi and Tripathi 1984), these counts are in good agreement with the 923 ganglion cells found in one eye in which the total number of such cells in the entire ciliary muscle was evaluated. This number is smaller than that of NOS/NADPH-

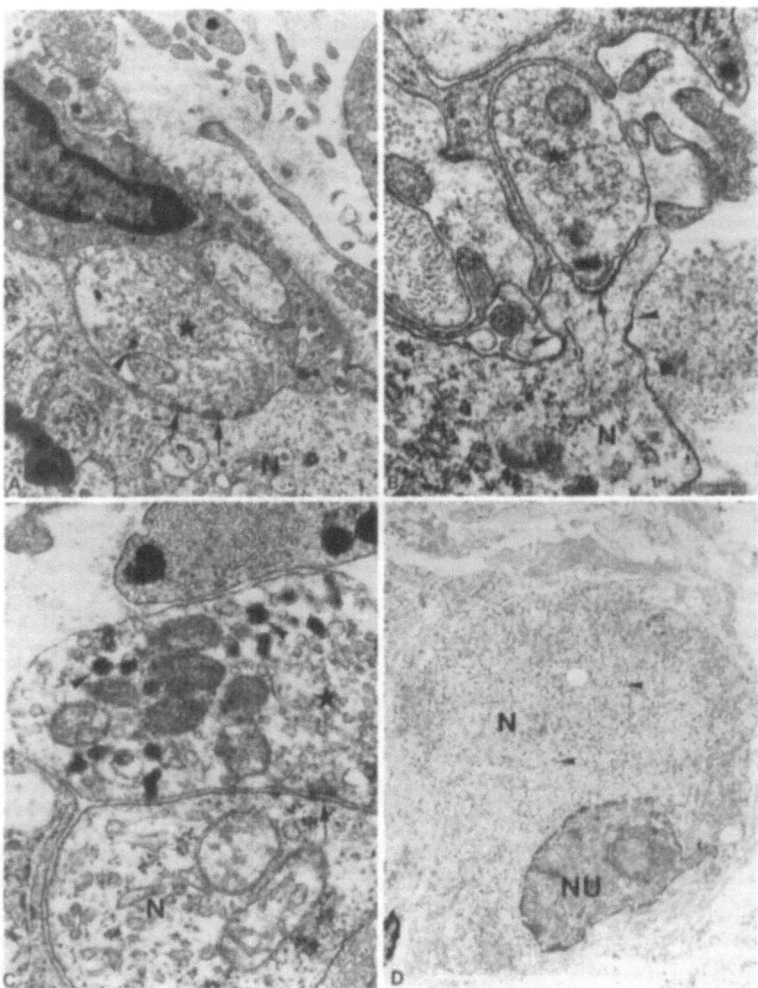

FIG. 5. Electron microscopy of ciliary muscle neurons (donor age 56 years). **A** Ciliary muscle nerve cells are in contact with terminal boutons that form axosomatic synaptic contacts (*arrows*). The boutons contain numerous agranular vesicles (40–60 nm) (*star*) and some large granular vesicles (60–120 nm) (*arrowheads*). (×50 000) (**B,C**) Ciliary muscle neurons (*N*) express spine-like (**B**, *arrowheads*) and lamellar (**C**) dendritic processes where axodendritic synapses (*arrow*) are formed. In addition to agranular vesicles (*star*), some of the axo-dendritic synaptic boutons contain a larger number of granular vesicles than do the axosomatic contacts (**C**, *arrowheads*). **D** Ultrahistochemical investigation of NADPH-D-stained ganglion cells (*N*) shows the electron-dense NADPH-D reaction product in the cytoplasm (*arrowheads*). The neurons show structural features similar to those observed by conventional electron microscopy. *NU*, nucleus. (×6000) (From Tamm et al. 1995a, with permission of Lippincott-Raven)

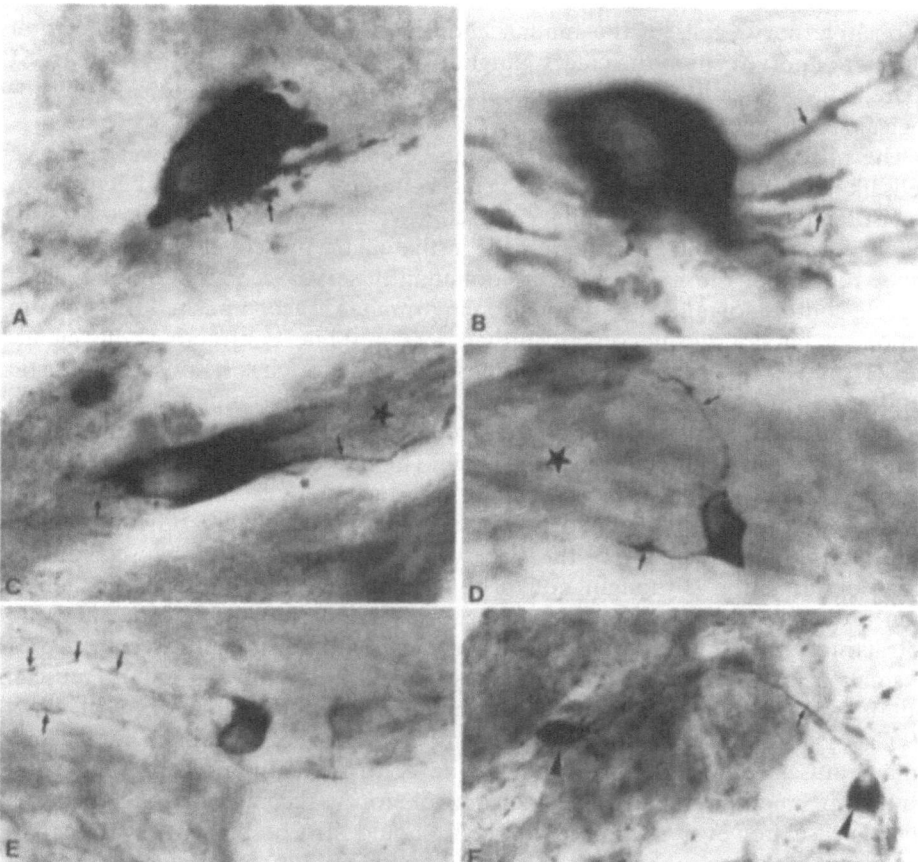

Fig. 6. NADPH-D-positive nerve cells in the ciliary muscle (tangential sections). **A,B** Large ciliary muscle ganglion cells show a prominent NADPH-D reaction. Both soma and dendrites of the nerve cells are darkly stained. The cells are multipolar and express numerous lamellar dendrites (**A**, *arrows*) or several long and short filamentous processes (**B**, *arrows*). (×1000) **C** Long filamentous processes of large neurons (*arrows*) run close and parallel to adjacent ciliary muscle bundles that show a faint NADPH-D reaction (*star*). (×1000) **D** A small ciliary muscle neuron expresses two long, filamentous processes (*arrows*) that encircle an adjacent muscle bundle (*star*). (×1000) **E** During their course along the muscle bundles, filamentous processes of ciliary muscle neurons express periodic swellings, suggesting varicosities (*arrows*). (×1000) **F** Two neighboring nerve cells (*arrowheads*) are in contact with each other by axonal processes (*arrows*). (×250) (From Tamm et al. 1995a, with permission of Lippincott-Raven)

D-positive nerve cells in the human choroid (Flügel et al. 1994) or the total number of nerve cells in the ciliary ganglion (1088–6835, mean ± SD 2394 ± 1036) (Perez and Keyser 1986), but it appears to be large enough to indicate physiological significance.

The nitrergic innervation of ciliary muscle cells via intrinsic nitrergic nerve cells likely induces relaxation. In vitro an NO-induced relaxation has been shown for isolated bovine and cat ciliary muscle (Wiederholt et al. 1994; Goh et al. 1995). Most of the nitrergic nerve cells in the ciliary muscle are situated in the inner regions of the muscle that predominantly serve accommodation. Ciliary muscle contraction and accommodation in primates is accomplished through the action of cholinergic nerves (Kaufman 1992). While contracting, the ciliary muscle moves in an anterior–inward position and stretches its posterior elastic tendons (Rohen 1952, 1964; Tamm et al. 1991b). After accommodation, during relaxation and disaccommodation the stretched posterior elastic tendons retract and pull the ciliary muscle backward again. The nerve fibers that mediate ciliary muscle contraction have their origin in the nucleus of Edinger-Westphal and synapse in the ciliary ganglion (Warwick 1954; Ruskell and Griffiths 1979; Kaufman 1992). In contrast, the backward movement of the muscle during disaccommodation is thought not to be mediated by an antagonistic innervation but to be caused only by the decrease of the cholinergic input and the retracting force of the muscle's posterior elastic tendons. There is evidence of a small relaxing effect of the adrenergic innervation of the ciliary muscle, but it is probably without major physiological significance (Törnqvist 1966, 1967a). The presence of intrinsic NO-synthesizing ganglion cells in the human ciliary muscle might therefore indicate, in contrast to the current belief, that a relaxing innervation plays an important role for disaccommodation in general. In this case, however, such ganglion cells should also be present in the ciliary muscle of higher monkeys, the only species with a ciliary muscle system almost similar in structure and function to that in humans (Rohen 1962; Kaufman 1992). In the eyes of most lower mammals, the ciliary muscle is weakly developed or vestigial, and the accommodative amplitude is much smaller than that of the human or monkey (Franz 1934; Duke-Elder 1958; Bito and Miranda 1987, 1989).

To obtain further information about the role of NOS-positive ganglion cells in the human ciliary muscle, we looked for similar ganglion cells in the ciliary muscle system of two monkey species: cynomolgus monkeys (*Macaca fascicularis*) and owl monkeys (*Aotes trivirgatus*). Both of these species have a well-developed ciliary muscle and an accommodative amplitude comparable to that of humans (Rohen 1962; Törnqvist 1967a, 1967b; Chin et al. 1968). We found numerous NOS/NADPH-D-positive ganglion cells in the ciliary muscle of cynomolgus monkeys, although they appeared to be less dense than in human ciliary muscle (Tamm and Lütjen-Drecoll 1997). In contrast, no such nerve cells were found in the ciliary muscle of owl monkeys, indicating that nitrergic nerve cells are not necessary for disaccommodation of primate ciliary muscle in general. Humans and cynomolgus monkeys are diurnal with a well-developed fovea centralis. Owl monkeys are nocturnal animals and do not have a fovea (Rohen 1962). Nitrergic ganglion cells in the ciliary muscle might therefore play a role in the fine adjustment of ciliary muscle contraction in species with high requirements for visual acuity. They might smooth the ciliary ganglion-mediated contraction or contribute to the small fluctuations or oscillations of accommodation observed under steady viewing conditions (Campbell et al. 1959). These fluctuations are assumed to help indicate the direction (contraction or relaxation) in which change should occur to obtain perfect focus.

Double-labeling experiments showed that approximately 70% of nitrergic ciliary muscle ganglion cells were in contact with varicose nerve endings expressing substance P (SP)-like and calcitonin gene-related peptide (CGRP)-like immunoreactivity (Tamm et al. 1995a) (Fig. 7). The origin of these nerves is not clear. We did not observe SP-IR or CGRP-IR perikarya in the human ciliary ganglion (Kirch et al. 1995). Some SP-IR but no CGRP-IR nerve cells have been demonstrated in the Edinger-Westphal nucleus of cats and cynomolgus monkeys and may also be present in humans (Maciewicz et al. 1983; Zhang et al. 1994a). On the other hand, ocular SP-IR and CGRP-IR axons have been shown to derive from sensory trigeminal ganglion cells in a considerable number of species (Stone et al. 1987). Such cells are also found in the ophthalmic portion of the human trigeminal ganglion (Quartu et al. 1992; Kirch et al. 1995). It is assumed that SP and CGRP are locally released by means of an axon reflex during the course of an ocular irritative response (Unger 1989; Bill 1991). Collaterals of SP- or CGRP-containing neurons that innervate ciliary muscle nerve cells may promote relaxation of the ciliary muscle in such events. Uveoscleral outflow is increased in cynomolgus monkeys after experimental iridocyclitis (Toris and Pederson 1987). Release of NO might contribute to this phenomenon because relaxation of the ciliary muscle and widening of the intermuscular spaces is known to cause a significant increase in uveoscleral outflow (Bill 1967). There is some evidence that the trigeminal system influences ciliary muscle tone also via the cholinergic nerve cells in the ciliary ganglion. In humans SP-IR axons closely surrounded 18.0% ± 1.5% of nerve cells in the ciliary ganglion, whereas CGRP-IR axons were in contact with 12.5% ± 0.7% of neurons (Kirch et al. 1995). In double-labeling experiments, almost all CGRP-IR axons were also SP-IR. It is tempting to speculate that contacts of trigeminal SP/CGRP axon collaterals with cholinergic nerve cells in the ciliary ganglion (involved in ciliary muscle contraction) and with nitrergic nerve cells in the ciliary muscle (involved in ciliary muscle relaxation) might function as alternative pathways to reflex arcs through the brain stem, which enables trigeminal neurons to modulate (increase or decrease) ciliary muscle tone (Fig. 8).

The NADPH-D reaction stains not only nerve cells in the ciliary muscle but also, albeit with weaker intensity, ciliary muscle cells themselves (Nathanson and McKee 1995a; Tamm et al. 1995a). It is likely that this staining does not reflect the presence of neuronal NOS, as antibodies against this isoform do not react with ciliary muscle cells (Nathanson and McKee 1995a; Tamm et al. 1995a). Immunohistochemical findings by Nathanson and McKee (1995a) indicate that

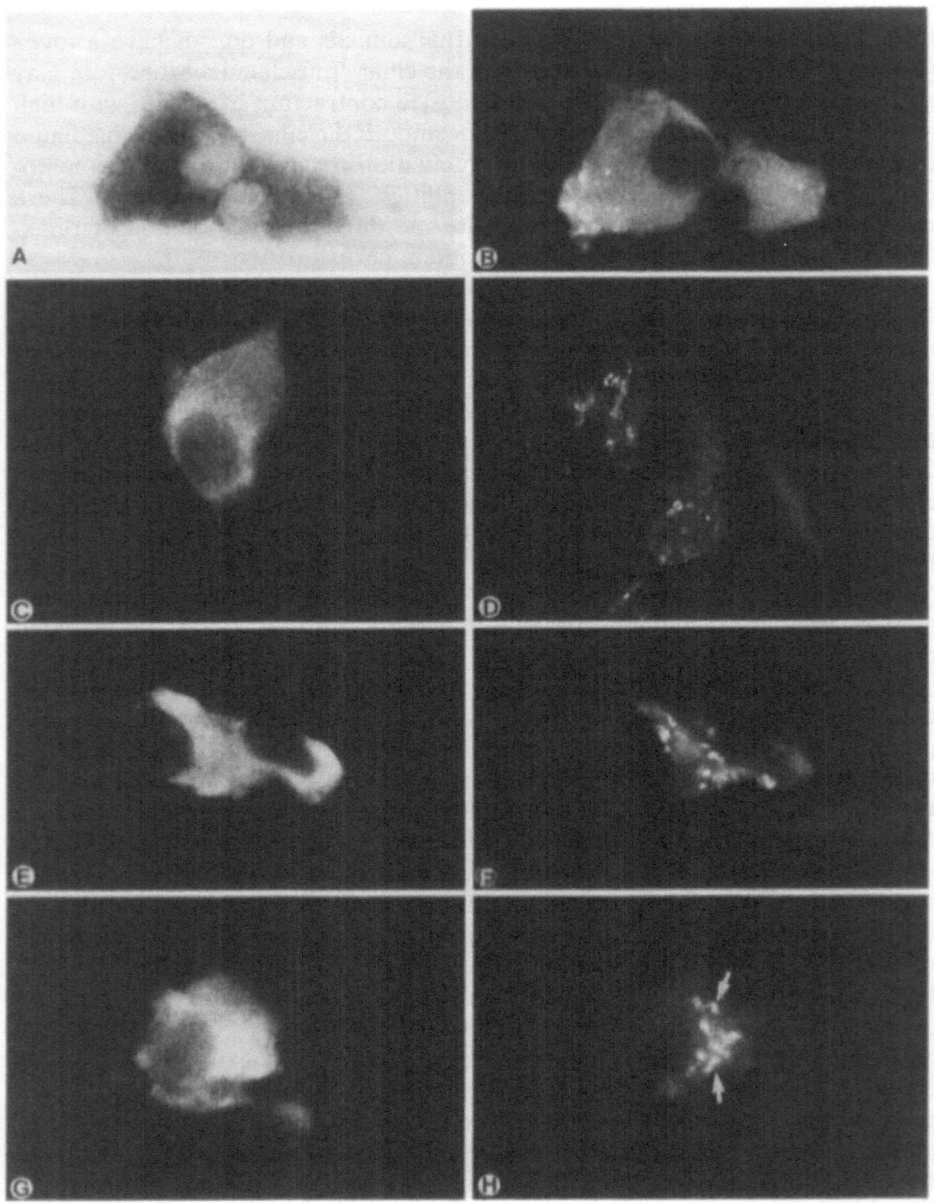

Fig. 7. Immunocytochemistry of ciliary muscle neurons. **A,B** Combination of NOS immunostaining (**B**) and NADPH-D (**A**) reveals complete co-localization in neuronal perikarya and fibers. (×1800) **C,D** Double labeling of ciliary muscle neuron for NOS (**C**) and substance P (**D**). Substance P-like immunoreactivity is seen in varicose axons that encircle the NOS-immunoreactive neuron. (×2000) **E–H** Large (**E,F**: ×1800) and small (**G,H**: ×2000) ciliary muscle neurons that stain for NOS (**E,G**) are in contact with arborizing boutons that express calcitionin gene-related peptide (CGRP)-like immuno-reactivity (*arrows*). (From Tamm et al. 1995a, with permission of Lippincott-Raven)

Fig. 8. Pathways of a presumptive reflex arc between sensory ophthalmic nerve cells (*arrow*) in the trigeminal ganglion and cholinergic (ciliary muscle contracting) nerve cells (*arrowhead*) in the ciliary ganglion. A similar pathway could exist for nitrergic (ciliary muscle relaxing) intrinsic nerve cells in the ciliary muscle

endothelial nitric oxide synthase (eNOS) in ciliary muscle cells might be responsible for the positive NADPH-D staining, although some unspecific activity of the antibodies used to detect eNOS was noted by the authors. eNOS immunoreactivity has been found in mitochondria isolated from most mammalian tissues (Bates et al. 1996) and is especially abundant in oxidative, mitochondria-rich skeletal muscle fibers (Kobzik et al. 1995). Mitochondria may also contribute to positive eNOS and NADPH-D staining in ciliary muscle cells because, as mentioned, primate ciliary muscle cells are rich in mitochondria (Ishikawa 1962; van der Zypen 1967; Flügel et al. 1990; Tamm and Lütjen-Drecoll 1996b). In addition, many structural details of ciliary muscle cells, such as the almost parallel myofibrils, Z-band-like extensions of the dense bands, rich innervation, and absence of gap junctions, resemble those of striated skeletal (Ishikawa 1962; van der Zypen 1967; Samuel et al. 1996; Tamm and Lütjen-Drecoll 1996b), as does the ciliary muscle's ability to contract and relax rapidly (Kaufman 1992). The role for mitochondrial NOS must be determined; two possibilities exist: an involvement in oxidative phosphorylation and the inactivation of mitochondrialderived oxygen free radicals (Kobzik et al. 1995; Bates et al. 1996). Nathansson and McKee (1995b) reported that NADPH-D-staining in the ciliary muscle of eyes with primary open-angle glaucoma (POAG) was less intense than in controls. Although it is emphasized that the NADPH-D reaction is a qualitative, not a quantitative method, a decrease in muscular NADPH-D staining may indicate that structural changes in ciliary muscle cells are associated with POAG. Such changes have been reported (Fine et al. 1981), but it is likely that they are symptoms rather than a cause of POAG, as they are also found in eyes with secondary glaucoma (Ueno and Naumann 1989). Nathanson and McKee (1995b) also reported that in POAG eyes the anterior insertion of the ciliary muscle to the scleral spur and trabecular meshwork was more posterior than in controls and suggested a glaucoma-associated loss of anterior ciliary muscle cells. These findings are difficult to interpret, as large interindividual variations exist regarding the position of the anterior insertion of the ciliary muscle in the normal human eye (Fischer 1933).

In addition to ciliary muscle cells, NADPH-D activity was observed in the non-pigmented layer of the human ciliary epithelium (Nathanson and McKee 1995a), corroborating earlier findings in the rabbit (Osborne et al. 1993). This activity is observed following formaldehyde fixation, but it is especially pronounced in unfixed tissues (authors' observations). Because in nonneuronal tissues the NADPH-D reaction is sensitive but not specific for detecting NOS activity, especially in unfixed tissues (Matsumoto et al. 1993; Tracey et al. 1993), it is at present not clear to what extent NADPH-D staining of the ciliary epithelium reflects the presence of an isoform of NOS. NOS appears to be present in other secretory epithelia, including the epithelium of the choroid plexus in the brain (Kobzik et al. 1993; Lin et al. 1996; Price et al. 1996; Szmydynger-Chodobska et al. 1996), which like the ciliary epithelium is an actively secreting epithelium of neuroec-todermal origin. The functional role for NO in secreting epithelia is not clear. In the feline small intestine, NOS inhibition increased epithelial permeability (Kubes 1992). Positive NADPH-D reactivity was also observed in the human tra-becular meshwork by Nathanson and McKee (1995a) but was not observed in our studies, at least not with incubation times sufficient to stain nerve cells, ciliary muscle, and ciliary epithelial cells (Fig. 9). This indicates that NADPH-D (and

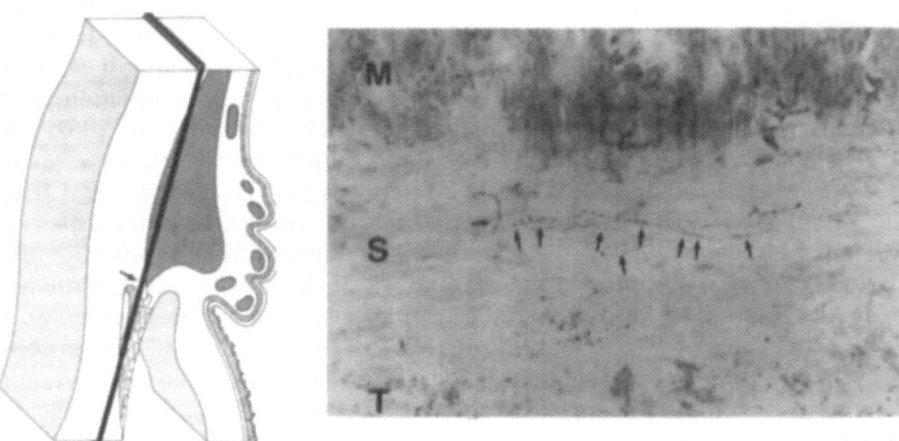

FIG. 9. Tangential section through the scleral spur region in a human eye (*drawing* shows plane of section; *arrow* indicates scleral spur). The NADPH-D-reaction labels numerous varicose, circumferentially oriented axons (*arrows*) in the scleral spur (*S*). Longitudinal ciliary muscle bundles (*M*) at their insertion to the scleral spur are seen at the top of the micrograph. The muscle cells are also stained by the NADPH-D reaction. No such staining is seen in the trabecular meshwork (*T*), which is seen at the bottom of the micrograph. (×300)

presumably NOS) activity in human trabecular meshwork is considerably weaker than in other cell types of the human anterior eye segment.

Nitrergic Nerves and Intraocular Pressure

In human eyes, numerous circumferentially oriented NOS/NADPH-D-positive varicose axons are found in the region of the scleral spur, a wedge-shaped ridge that projects from the inner side of the anterior sclera (Tamm et al. 1995b). At its anterior aspect, the scleral spur provides an attachment for the connective tissue elements of the corneoscleral and cribriform parts of the trabecular meshwork; on its posterior side, the anterior tendons of the ciliary muscle attach (Rohen 1956b; Kupfer 1962).

The resident cells within the scleral spur (scleral spur cells; SSCs) show structural characteristics of contractile myofibroblasts (Tamm et al. 1992). They stain for smooth muscle α-actin and myosin but not for desmin, which is the characteristic intermediate filament of ciliary muscle cells (Tamm et al. 1991a). In contrast to the adjacent longitudinal ciliary muscle bundles, the SSCs form no bundles and are circumferentially oriented. They do form tendon-like contacts with the elastic fibers of the scleral spur, which are continuous with those of the adjacent trabecular meshwork. It seems reasonable to assume that changes in SSC tone cause changes in the trabecular meshwork architecture, thereby modulating the size of the outflow pathways for aqueous humor.

Individual SSCs are innervated by autonomic nerve endings that closely contact their cell membrane (Tamm et al. 1992). We characterized the nature of this innervation and found that a considerable number of the nerve fibers contacting SSCs stained for NOS and NADPH-D (Fig. 9). Other nerve fibers were immunoreactive for SP, CGRP, vasointestinal peptide (VIP), and neuropeptide Y (NPY) (Tamm et al. 1995b). The nitrergic axons in the scleral spur derive from nerve fiber bundles that run in the supraciliary space and appear to come from the posterior eye segment. They probably originate from the pterygopalatine ganglion or perhaps from choroidal ganglion cells. Thus far the physiological role of the nitrergic SSC innervation must remain a subject of speculation as no experimental data on the action of SSCs in human eyes are available. It seems reasonable to assume that a nitrergic innervation causes a decrease in SSC tone.

In isolated bovine trabecular meshwork strips, relaxation was induced by application of organic nitrovasodilators (Wiederholt et al. 1994). The posterior part of the bovine trabecular meshwork consists of numerous myofibroblast-like cells and appears to be an equivalent to the scleral spur region in the human eye (Flügel et al. 1991). Agents that cause relaxation of isolated bovine meshwork increase outflow facility in organ cultures of bovine anterior segments (Wiederholt et al. 1996). In human and monkey eyes, application of organic nitrovasodilators decreases intraocular pressure (IOP), an effect that in monkeys is caused by a decrease in resistance to aqueous humor outflow (Wizemann and

Wizemann 1980; Schuman et al. 1994). In summary, then, it seems likely that NO released from nitrergic scleral spur nerves decreases outflow resistance in the human eye.

After having passed through the trabecular meshwork into Schlemm's canal, aqueous humor leaves the eye via flat collector channels that lead into the episcleral venous plexus (Ashton 1952; Jocson and Grant 1965; Jocson and Sears 1968, 1969; Funk and Rohen 1995). Resistance to aqueous humor outflow depends on the pressure difference between the IOP and the episcleral venous pressure. There is some evidence that nitrergic nerves might influence outflow resistance not only by modulating resistance in the trabecular meshwork but also by changing the diameter of the episcleral vessels, thereby influencing episcleral venous pressure. In the human eye the individual vessels of the episcleral venous plexus are in contact with a dense network of NADPH-D/NOS-positive axons (Fig. 10). In addition, their endothelial cells stain intensely for NADPH-D, suggesting the presence of eNOS. In the rabbit, dog, monkey, and humans the episcleral venous plexus forms arteriovenous anastomoses with arterioles that derive from branches of the anterior ciliary arteries (Meighan 1956; Rohen 1956a; Gaasterland et al. 1970; Rohen and Funk 1994; Funk and Rohen 1995). There is some evidence that in the rabbit the diameter of the anastomoses and the amount of blood shunted may influence episcleral venous pressure and, indirectly, outflow resistance (Funk et al. 1996). The arteriolar part of the arteriovenous anastomoses in rat, rabbit, and dog exhibits dense nitrergic innervation (Funk et al. 1994).

In summary, there seem to be two mechanisms by which nitrergic nerves might modulate IOP: (1) changing the tone of scleral spur cells, thereby modulating the architecture of the trabecular meshwork and in turn its outflow resistance; and (2) dilating episcleral vessels, thereby lowering episcleral venous pressure.

Nitrergic Nerves in the Eyelid

In a considerable number of mammalian species, including the cat, rabbit, rat, and monkey, a contribution of postganglionic pterygopalatine nerves to conjunctival and upper lid innervation has been demonstrated by nerve degeneration and retrograde tracing experiments (Macintosh 1974; Uddman et al. 1980; Butler et al. 1984; van der Werf 1993; Elsas et al. 1994). Because most neurons in the pterygopalatine ganglion show NOS and NADPH-D reactivity, it seemed likely that nitrergic nerves are involved in conjunctival and upper lid innervation.

We studied the nitrergic innervation of the upper lid in the cynomolgus monkey and found a dense network of nitrergic axons associated with the meibomian (tarsal) glands (Kirch et al. 1996). NOS/NADPH-D-positive axons curved in close contact along the basal side of the basal glandular cell layer and the epithelial cells lining the central duct of the glands. Nitrergic axons were

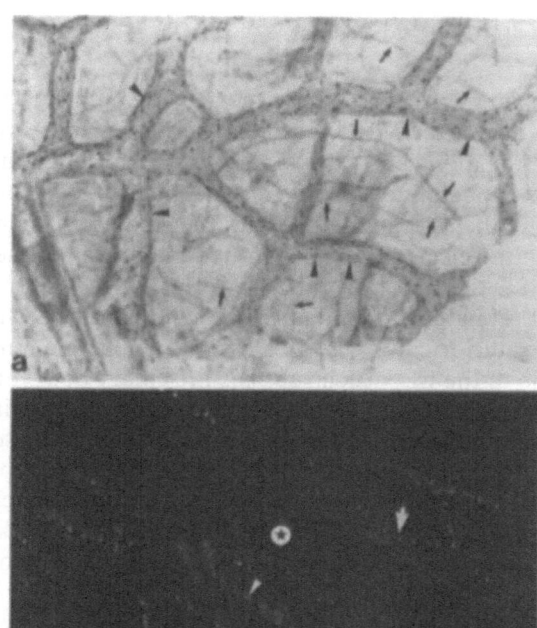

FIG. 10. Tangential sections through the episcleral region in a human eye. **a** Dense network of NADPH-D-positive axons (*arrows*) is observed in close contact with the vessels forming the episcleral venous plexus (*arrowheads*). In addition, there is marked positive NADPH-D staining in the endothelial cells of the episcleral vessels. (×200) **b** Episcleral arteriole (*arrowheads*) and an associated venule (*star*) are both surrounded by a dense perivascular network of NOS-positive axons (*arrows*). (×300)

also observed in the wall of small blood vessels near meibomian gland acini. In addition to NOS- and NADPH-D-positive axons, meibomian glands were in close contact with axons showing immunoreactivity for SP, CGRP, VIP, NPY, and, albeit more rarely, the adrenergic markers tyrosine hydroxylase and dopamine-β-hydroxylase. Meibomian glands secrete lipid that is delivered to the anterior face of the tear film. The main function of meibomian lipid is to reduce evaporation from the corneal surface and contribute to the stability of the precorneal tear film (Tiffany 1995). So far, the role of nitrergic nerves for meibomian gland function is not clear, but stimulation of secretion might be possible.

Conclusion

In the human eye nitrergic perivascular nerves are numerous in the wall of choroidal arteries and arterioles. Choroidal perivascular nerves derive from intrinsic choroidal ganglion cells (Fig. 11a) or extrinsic ganglion cells, probably localized in the pterygopalatine and trigeminal ganglia. Axons of these extrinsic nerve cells penetrate the sclera together with the small posterior ciliary arteries (Fig. 11b). Similar nitrergic axons are also found in the wall of the central retinal artery (Fig. 11c) and the anterior ciliary arteries (Fig. 11d). Nitrergic nerves around the central retinal artery appear not to enter the eye, whereas nitrergic axons around the anterior ciliary arteries form the nitrergic innervation of the major arterial circle of the iris, which supplies the ciliary body and iris. Nitrergic nerve cells in the ciliary muscle (Fig. 11e) appear to innervate ciliary muscle cells and to contribute to disaccommodation. Nitrergic nerves in the scleral spur (Fig. 11f), close to the trabecular meshwork, and around the episcleral vessels (Fig. 11g) might modulate IOP by (1) changing the tone of scleral spur cells, thereby modulating the architecture of the trabecular meshwork and in turn its outflow resistance; and (2) dilating episcleral vessels, thereby lowering episcleral venous pressure.

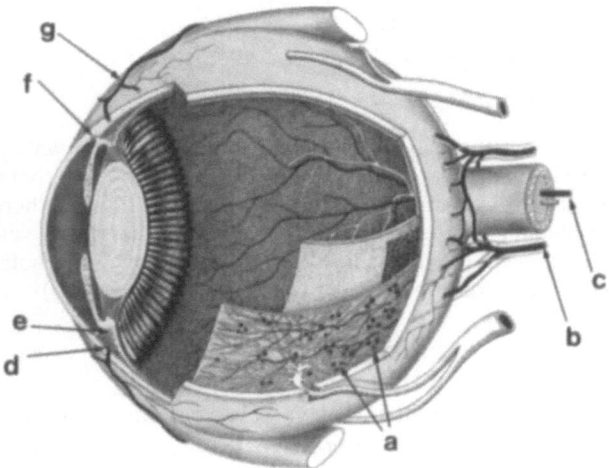

FIG. 11. For legend see Conclusion

Acknowledgments. This work was supported by grants from the Ria Freifrau von Fritsch-Stiftung of the University Erlangen-Nürnberg, Germany (ERT) and the Deutsche Forschungsgemeinschaft (ELD, Dr 124/6-3). ERT is the recipient of a Heisenberg Award from the Deutsche Forschungsgemeinschaft (Ta 115/8-1). We thank Drs. Joram Piatigorsky, Paul L. Kaufman, and Paul Russell for their critical reading of this manuscript.

References

Alder VA, Cringle SJ, Constable IJ (1983) The retinal oxygen profile in cats. Invest Ophthalmol Vis Sci 24:30–36

Alm A (1992) Ocular circulation. In: Hart WMJ (ed) Adler's physiology of the eye. Mosby-Year Book, St. Louis, pp 198–227

Alm A, Bill A (1970) Blood flow and oxygen extraction in the cat uvea at normal and high intraocular pressures. Acta Physiol Scand 80:19–28

Alm A, Bill A (1972) The oxygen supply to the retina. II. Effects of high intraocular pressure and of increased arterial carbon dioxide tension on uveal and retinal blood flow in cats: a study with radioactively labelled microspheres including flow determinations in brain and some other tissues. Acta Physiol Scand 84:306–319

Alm A, Bill A (1973) Ocular and optic nerve blood flow at normal and increased intraocular pressures in monkeys (*Macaca irus*): a study with radioactively labelled microspheres including flow determinations in brain and some other tissues. Exp Eye Res 15:15–29

Ashton N (1952) Anatomical study of Schlemm's canal and aqueous veins by means of neoprene casts. II. Aqueous veins. Br J Ophthalmol 36:265

Bates TE, Loesch A, Burnstock G, Clark JB (1996) Mitochondrial nitric oxide synthase: a ubiquitous regulator of oxidative phosphorylation? Biochem Biophys Res Commun 218:40–44

Beesley JE (1995) Histochemical methods for detecting nitric oxide synthase. Histochem J 27:757–769

Bergua A (1995) Nitrergische Reaktivität in den Endothelzellen der menschlichen uvealen Gefässe. Klin Monatsbl Augenheilkd 206:115–121

Bergua A (1996) NADPH-diaphorase-positive innervation of the central retinal artery of the human optic nerve. Exp Eye Res 63(Suppl):S.142

Bergua A, Jünnemann A, Naumann GOH (1993) NADPH-D reactive choroidal ganglion cells in the human eye. Klin Monatsbl Augenheilkd 203:77–82

Bergua A, Neuhuber WL, Naumann GOH (1994) Visualization of human choroidal ganglion cells with the supravital fluorescent dye 4-(4-diethylaminostyryl)-N-methylpyridium iodide. Ophthalmic Res 26:290–295

Bergua A, Neuhuber WL, Mayer B (1995) Comparative anatomy of nitrinergic innervation in avian choroid. Invest Ophthalmol Vis Sci 36:S121 (ARVO abstracts)

Bergua A, Mayer B, Neuhuber WL (1996) Nitrergic and VIPergic neurons in the choroid and ciliary ganglion of the duck Anis carina. Anat Embryol (Berl) 193:239–248

Bill A (1967) Effects of atropine and pilocarpine on aqueous humor dynamics in cynomolgus monkeys (*Macaca irus*). Exp Eye Res 6:120–125

Bill A (1968) Capillary permeability to and extravascular dynamics of myoglobin, albumin and gammaglobulin in the uvea. Acta Physiol Scand 73:204–219

Bill A (1985) Some aspects of the ocular circulation: Friedenwald lecture. Invest Ophthalmol Vis Sci 26:410–424

Bill A (1991) The 1990 Endre Balazs lecture: effects of some neuropeptides on the uvea. Exp Eye Res 53:3–11

Bill A, Törnquist P, Alm A (1980) The permeability of the ocular vessels. Trans Ophthalmol Soc UK 100:332–336

Bito LZ, Miranda OC (1987) On the evolution of visual accommodation, the non-accommodating eye and presbyopia. In: De Vincentiis M (ed) The fundamental aging processes of the eye. Fondazione "Giorgio Ronchi" LX, Tipographia Baccini & Chiappi, Florence, pp 58–97

Bito LZ, Miranda OC (1989) Accommodation and presbyopia. In: Reinecke RD (ed) Ophthalmology annual. Lippincott-Raven, New York, pp 103–127

Blottner D, Grozdanovic Z, Gossrau R (1995) Histochemistry of nitric oxide synthase in the nervous system. Histochem J 27:785–811

Boeke J (1933) Innervationsstudien. III. Die Nervenversorgung des M. ciliaris und des sphincter iridis bei Säugern und Vögeln. Z Mikrosk Anat Forsch 33:233–275

Bredt DS, Hwang PM, Snyder SH (1992) Localization of nitric oxide indicating a neuronal role for nitric oxide. Nature 347:768–770

Bryson JM, Wolter JR, O'Keefe NT (1966) Ganglion cells in the human ciliary body. Arch Ophthalmol 75:57–60

Butler JM, Ruskell GL, Cole DF, Unger WG, Zhang SQ, Blank MA, McGregor GP, Bloom SR (1984) Effects of VIIth (facial) nerve degeneration on vasoactive intestinal polypeptide and substance P levels in ocular and orbital tissues of the rabbit. Exp Eye Res 39:523–532

Campbell FW, Robson JG, Westheimer G (1959) Fluctuations in accommodation under steady viewing conditions. J Physiol (Lond) 145:579–594

Castro-Correira J (1967) Studies on the innervation of the uveal tract. Ophthalmologica 154:497–520

Chakravarthy U, Stitt AW, McNally J, Bailie JR, Hoey EM, Duprex P (1994) Nitric oxide synthase activity and expression in retinal capillary endothelial cells and pericytes. Curr Eye Res 14:285–294

Chin NB, Ishikawa S, Lappin H, Davidowitz J, Breinin GM (1968) Accommodation in monkeys induced by midbrain stimulation. Invest Ophthalmol Vis Sci 7:386–396

Dawson TM, Bredt DS, Fotuhi PM, Hwang PM, Snyder SH (1991) Nitric oxide synthase and neuronal NADPH diaphorase are identical in brain and peripheral tissues. Proc Natl Acad Sci USA 88:7797–7801

De Stefano ME, Luzzatto AC, Mugnaini E (1993) Neuronal ultrastructure and somatostatin immunolocalization in the ciliary ganglion of chicken and quail. J Neurocytol 22:868–892

Deussen A, Sonntag M, Vogel R (1993) L-Arginine-derived nitric oxide: a major determinant of uveal blood flow. Exp Eye Res 57:129–134

Ding YQ, Takada M, Kaneko T, Mizuno M (1995) Colocalization of vasoactive intestinal polypeptide and nitric oxide in penis-innervating neurons in the major pelvic ganglion of the rat. Neuroreport 22:129–131

Donati G, Pournaras CJ, Munoz J-L, Poitry S, Poitry-Yamate CL, Tsacopoulos M (1995) Nitric oxide controls arteriolar tone in the retina of the miniature pig. Invest Ophthalmol Vis Sci 36:2228–2237

Dryer SE, Chiapinelli VA (1985) Properties of choroid and ciliary neurons in the avian ciliary ganglion and evidence for substance P as a neurotransmitter. J Neurosci 5:2654–2661

Duke-Elder S (1958) System of ophthalmology. Vol. I. The eye in evolution. Mosby, St. Louis

Ehinger B, Sundler F, Uddman R (1983) Functional morphology in two parasympathetic ganglia: the ciliary and the pterygopalatine. In: Elfvin L-G (ed) Autonomic ganglia. Wiley, New York, pp 97–123

Elsas T, Edvinsson L, Sundler F, Uddman R (1994) Neuronal pathways to the rat conjunctiva revealed by retrograde tracing and immunocytochemistry. Exp Eye Res 58:117–126

Feeney L, Hogan MJ (1961) Electron microscopy of the human choroid. II. The choroidal nerves. Am J Ophthalmol 51:1072–1083

Fine BS, Yanoff M, Stone RA (1981) A clinicopathologic study of four cases of primary open-angle glaucoma compared to normal eyes. Am J Ophthalmol 91:88–105

Fischer F (1933) Entwicklungsgeschichtliche und anatomische Studien über den Skleralsporn im menschlichen Auge. Graefes Arch Ophthalmol 133:318–358

Fitzgerald MEC, Reiner A (1993) NADPH-diaphorase positive neurons and fibers in the ciliary ganglion and choroid of the pigeon. Soc Neurosci Abstr 19:1202 (abstract)

Flügel C, Bárány EH, Lütjen-Drecoll E (1990) Histochemical differences within the ciliary muscle and its function in accommodation. Exp Eye Res 50:219–226

Flügel C, Tamm E, Lütjen-Drecoll E (1991) Different cell populations in bovine trabecular meshwork: an ultrastructural and immunohistochemical study. Exp Eye Res 52:681–690

Flügel C, Tamm ER, Mayer B, Lütjen-Drecoll E (1994) Species differences in choroidal vasodilative innervation: evidence for specific intrinsic nitrergic and VIP-positive neurons in the human eye. Invest Ophthalmol Vis Sci 35:592–599

Flügel-Koch C, Kaufman PL, Lütjen-Drecoll E (1994) Association of choroidal ganglion cell plexus with the fovea centralis. Invest Ophthalmol Vis Sci 35:4268–4272

Flügel-Koch C, May CA, Lütjen-Drecoll E (1996) Presence of a contractile cell network in the human choroid. Ophthalmologica 210:296–302

Frank RN, Turczyn TJ, Das A (1990) Pericyte coverage of retinal and cerebral arteries. Invest Ophthalmol Vis Sci 31:999–1007

Franz V (1934) Vergleichende Anatomie des Wirbeltierauges. In: Bolk L, Göppert E, Kallius E, Lubasch W (eds) Handbuch der vergleichenden Anatomie der Wirbeltiere. Vol. II/2. Urban & Schwarzenberg, Berlin, pp 1–1202

Funk R (1991) Ultrastructure of the ciliary process vasculature in cynomolgus monkeys. Exp Eye Res 53:461–469

Funk R, Rohen JW (1988) SEM studies of the functional morphology of the ciliary process vasculature in the cynomolgus monkey: reactions after application of epinephrine. Exp Eye Res 47:653–663

Funk R, Rohen JW (1990) Scanning electron microscopic study on the vasculature of the human anterior eye segment, especially with respect to the ciliary processes. Exp Eye Res 51:651–661

Funk RHW, Rohen JW (1995) Scanning electron microscopic study of episcleral arteriovenous anastomoses in the owl and cynomolgus monkey. Curr Eye Res 15: 321–327

Funk RHW, Mayer B, Wörl J (1994) Nitrergic innervation and nitrergic cells in arteriovenous anastomoses. Cell Tissue Res 277:477–484

Funk RHW, Gehr J, Rohen JW (1996) Short-term hemodynamic changes in episcleral arteriovenous anastomoses correlate with venous pressure and IOP changes in the albino rabbit. Curr Eye Res 15:87–93

Furness JB, Bornstein JC, Trussell DC (1988) Shapes of nerve cells in the myenteric plexus of the guinea-pig small intestine revealed by the intracellular injection of dye. Cell Tissue Res 254:561–571

Gaasterland DE, Jocson VL, Sears ML (1970) Channels of aqueous outflow and related blood vessels. II. Episcleral arteriovenous anastomoses in the rhesus monkey eye (*Macaca mulatta*). Arch Ophthalmol 84:770–775

Goh Y, Hotehama Y, Mishima HK (1995) Characterization of ciliary muscle relaxation by various agents in cats. Invest Ophthalmol Vis Sci 36:1188–1192

Haefliger IO, Zschauer A, Anderson DR (1994) Relaxation of retinal pericyte contractile tone through the nitric oxid-cyclic guanosine monophosphate pathway. Invest Ophthalmol Vis Sci 35:991–997

Hirano N (1941) Nervöse Innervation des Corpus ciliare des Menschen. Graefes Arch Ophthalmol 142:549–559

Hope BC, Michael GJ, Knigge KM, Vincent SR (1991) Neuronal NADPH diaphorase is a nitric oxide synthase. Proc Natl Acad Sci USA 88:2811–2814

Ishikawa T (1962) Fine structure of the human ciliary muscle. Invest Ophthalmol Vis Sci 1:587–608

Iwanoff A (1874) Der Uvealtractus. In: Graefe A, Saemisch T (eds) Handbuch der gesamten Augenheilkunde. Engelmann, Leipzig, pp 265–287

Jocson VL, Grant WM (1965) Interconnections of blood vessels and aqueous vessels in human eyes. Arch Ophthalmol 73:707–720

Jocson VL, Sears ML (1968) Channels of aqueous outflow and related blood vessels. I. *Macaca mulatta* (rhesus). Arch Ophthalmol 80:104–114

Jocson VL, Sears ML (1969) Channels of aqueous outflow and related blood vessels. II. *Cercopithecus ethiops* (Ethiopian green or green vervet). Arch Ophthalmol 81:244–253

Kaufman PL (1992) Accommodation and presbyopia. In: Hart WMJ (ed) Adler's physiology of the eye. Mosby-Year Book, St. Louis, pp 391–411

Kirch W, Neuhuber W, Tamm ER (1995) Immunohistochemical localization of neuropeptides in the human ciliary ganglion. Brain Res 681:229–234

Kirch W, Horneber M, Tamm ER (1996) Characterization of meibomian gland innervation in the cynomolgus monkey (*Macaca fascicularis*). Anat Embryol (Berl) 193:365–375

Klatt P, Heinzel B, John M, Kastner M, Böhme E, Mayer B (1992) Ca^{2+}/calmodulin-dependent cytochrome-c reductase activity of brain nitric oxide synthase. J Biol Chem 267:11374–11378

Kobzik L, Bredt DS, Lowenstein CJ, Drazen J, Gaston B, Sugarbaker D, Stamler JS (1993) Nitric oxide synthase in human and rat lung: immunocytochemical and histochemical localization. Am J Respir Cell Mol Biol 9:371–377

Kobzik L, Stringer B, Balligand J-L, Reid MB, Stamler JS (1995) Endothelial type nitric oxide synthase in skeletal muscle fibers: mitochondrial relationships. Biochem Biophys Res Commun 211:375–381

Krause W (1861) Ganglienzellen im Orbiculus ciliaris. In: Krause W (ed) Anatomische Untersuchungen. Hannover, p 91

Krümmel H (1938) Die Nerven des menschlichen Ziliarkörpers. Graefes Arch Ophthalmol 138:845–865

Kubes P (1992) Nitric oxide modulates epithelial permeability in the feline small intestine. Am J Physiol 262:G1138-G1142

Kummer W, Fischer A, Mundel P, Mayer B, Hoba B, Philippin B, Preissler U (1992) Nitric-oxide synthase in VIP-containing vasodilator nerve-fibers in the guinea-pig. Neuroreport 3:653–655

Kupfer C (1962) Relationship of ciliary body meridional muscle and corneoscleral trabecular meshwork. Arch Ophthalmol 68:132–136

Kurus E (1955) Über ein Ganglienzellsystem der menschlichen Aderhaut. Klin Monatsbl Augenheilkd 127:198–206

Laties AM (1967) Central retinal artery innervation: absence of adrenergic innervation to intraocular branches. Arch Ophthalmol 77:405–409

Lauber H (1936a) Der Strahlenkörper (Corpus ciliare). D. Die Nerven des Strahlenkörpers. In: von Möllendorf W (ed) Handbuch der mikroskopischen Anatomie des Menschen. Vol 3: Haut und Sinnesorgane. Part 2: Auge. Springer, Berlin, pp 165–172

Lauber H (1936b) Die Aderhaut (Choroidea). In: von Möllendorf W (ed) Handbuch der mikroskopischen Anantomie. Vol 3: Haut und Sinnesorgane. Part 2: Auge. Springer, Berlin, pp 91–133

Lin AY-J, Szmydynger-Chodobska J, Rahman MP, Mayer B, Monfils PR, Johanson CE, Lim Y-P, Corsetti S, Chodobski A (1996) Immunohistochemical localization of nitric oxide synthase in rat anterior choroidal artery, stromal blood microvessels, and choroid plexus epithelial cells. Cell Tissue Res 285:411–418

Llewellyn-Smith IJ, Song ZM, Costa M, Bredt DS, Snyder SH (1992) Ultrastructural localization of nitric oxide synthase immunoreactivity in guinea pig enteric neurons. Brain Res 577:337–342

Lopez-Costa JJ, Goldstein J, Mallo G, Saavedra JP (1995) NADPH-diaphorase distribution in the choroid after continuous illumination. Neuroreport 26:361–364

Lou HC, Edvinsson L, MacKenzie ET (1987) The concept of coupling blood flow to brain function: revision required? Ann Neurol 22:289–297

Maciewicz R, Phipps BS, Foote WE, Aronin N, DiFiglia M (1983) The distribution of substance P-containing neurons in the cat Edinger-Westphal nucleus: relationship to efferent projection systems. Brain Res 270:217–230

Macintosh SR (1974) The innervation of the conjunctiva in monkeys: an electron microscopic and nerve degeneration study. Graefes Arch Klin Exp Ophthalmol 192:105–116

Mann RM, Riva CE, Stone RA, Barnes GE, Cranstoun SD (1995) Nitric oxide and choroidal blood flow regulation. Invest Ophthalmol Vis Sci 36:925–930

Matsumoto T, Nakane M, Pollock JS, Kuk JE, Förstermann U (1993) A correlation between soluble brain nitric oxide synthase and NADPH-diaphorase activity is only seen after exposure of the tissue to fixative. Neurosci Lett 155:61–64

May PJ, Warren S (1993) Ultrastructure of the macaque ciliary ganglion. J Neurocytol 22:1073–1095

Meighan SS (1956) Blood vessels of the bulbar conjunctiva in man. Br J Ophthalmol 40:513–526

Miller AS, Coster DJ, Costa M, Furness JB (1983) Vasoactive intestinal polypeptide immunoreactive nerve fibres in the human eye. Aust J Ophthalmol 11:185–193

Morris JL, Gibbins IL, Kadowitz PJ, Herzog H, Kreulen DL, Toda N, Claing A (1995) Roles of peptides and other substances in cotransmission from vascular autonomic and sensory neurons. Can J Physiol Pharmacol 73:521–332

Morris R, Southam E, Gittins SR, Garthwaite J (1993) NADPH-diaphorase staining in autonomic and somatic cranial ganglia of the rat. Neuroreport 4:62–64

Morrison JC, Van Buskirk EM (1983) Anterior collateral circulation in the primate eye. Ophthalmology 90:707–715

Morrison JC, Van Buskirk EM (1984) Ciliary process microvasculature of the primate eye. Am J Ophthalmol 97:372–384

Mulder H, Uddman R, Moller K, Elsas T, Ekblad E, Alumets J, Sundler F (1995) Pituitary adenylate cyclase activating polypeptide is expressed in autonomic neurons. Regul Pept 59:121–128

Müller H (1859a) Über Ganglienzellen im Ziliarmuskel des Menschen. Verh Physik Med Ges Würzburg 10:107–110

Müller H (1859b) Über glatte Muskelfasern und Nervengeflechte der Choroidea im menschlichen Auge. Verh Physik Med Ges Würzburg 10:107–179

Nathanson JA, McKee M (1995a) Identification of an extensive system of nitric oxide-producing cells in the ciliary muscle and outflow pathway of the human eye. Invest Ophthalmol Vis Sci 36:1765–1773

Nathanson JA, McKee M (1995b) Alterations of ocular nitric oxide synthase in human glaucoma. Invest Ophthalmol Vis Sci 36:1774–1784

Nilsson SFE (1994) PACAP-27 and PACAP-38: vascular effects in the eye and some other tissues in the rabbit. Eur J Pharmacol 253:17–25

Nilsson SFE (1996) Nitric oxide as a mediator of parasympathetic vasodilation in ocular and extraocular tissues in the rabbit. Invest Ophthalmol Vis Sci 37:2110–2119

Nilsson SFE, Bill A (1984) Vasoactive intestinal polypeptide (VIP): effects in the eye and on regional blood flows. Acta Physiol Scand 121:385–392

Nilsson SFE, Linder J, Bill A (1985) Characteristics of uveal vasodilation produced by facial nerve stimulation in monkeys, cats and rabbits. Exp Eye Res 40:841–852

Nyborg NCB, Nielsen PJ (1994) Neurogenic nitric oxide accounts for the non-adrenergic non-cholinergic vasodilation in human posterior ciliary arteries. Invest Ophthalmol Vis Sci 34:1287 (ARVO abstracts)

Osborne NN, Barnett NL, Herrera AJ (1993) NADPH diaphorase localization and nitric oxide synthetase activity in the retina and anterior uvea of the rabbit eye. Brain Res 610:194–198

Palmer RMJ, Ferrige AG, Moncada S (1987) Nitric oxide release accounts for the biological activity of endothelium-derived relaxing factor. Nature 327:524–525

Parver LM, Auker C, Carpenter DO (1980) Choroidal blood flow as a heat dissipating mechanism in the macula. Am J Ophthalmol 89:641–646

Parver LM, Auker CR, Carpenter DO, Doyle T (1982) Choroidal blood flow. II. Reflexive control in the monkey. Arch Ophthalmol 100:1327–1330

Parver LM, Auker CR, Carpenter DO (1983) Choroidal blood flow. III. Reflexive control in human eyes. Arch Ophthalmol 101:1604–1606

Perez GM, Keyser RB (1986) Cell body counts in human ciliary ganglia. Invest Ophthalmol Vis Sci 27:1428–1431

Perez MT, Larsson B, Alm P, Andersson KE, Ehinger B (1995) Localisation of neuronal nitric oxide synthase-immunoreactivity in rat and rabbit retinas. Exp Brain Res 104:207–217

Price KJ, Hanson PJ, Whittle BJR (1996) Localization of constitutive isoforms of nitric oxide synthase in the gastric glandular mucosa of rats. Cell Tissue Res 285:157–163

Quartu M, Diaz G, Floris A, Lai ML, Priestley JV, Del Fiacco M (1992) Calcitonin gene-related peptide in the human trigeminal sensory system at developmental and adult life stages: immunohistochemistry, neuronal morphometry and coexistence with substance P. J Chem Neuroanat 5:143–157

Riva CE, Harino S, Shonat RD, Petrig BL (1991) Flicker evoked increase in optic nerve head blood flow in anesthetized cats. Neurosci Lett 128:291–296

Rohen H (1952) Der Ziliarkörper als funktionelles System. Gegenbaur Morphol Jahrb 92:415–440

Rohen J (1956a) Arteriovenöse Anastomosen im Limbusbereich des Hundes. Graefes Arch Ophthalmol 157:361–367

Rohen JW (1956b) Über den Ansatz der Ciliarmuskulatur im Bereich des Kammer-winkels. Ophthalmologica 131:51–59

Rohen JW (1962) Sehorgan. In: Hofer H, Schultz AH, Starck D (eds) Primatologica, handbook of primatology. Vol. II/1. Karger, Basel, pp 6/1–210

Rohen JW (1964) Ciliarkörper (Corpus ciliare). In: von Möllendorf W, Bargmann W (eds) Handbuch der mikroskopischen Anatomie des Menschen. Vol 3. Part 4: Haut und Sinnesorgane. Das Auge und seine Hilfsorgane. Springer, Berlin Heidelberg New York, pp 189–237

Rohen JW (1982) The evolution of the primate eye in relation to the problem of glaucoma. In: Lütjen-Drecoll E (ed) Basic aspects of glaucoma research. Schattauer, Stuttgart, pp 3–33

Rohen JW, Funk RHW (1994) Functional morphology of the episcleral vasculature in the rabbit and canine eye: presence of arteriovenous anastomoses. J Glaucoma 3:51–57

Roufail E, Stringer M, Rees S (1995) Nitric oxide synthase immunoreactivity and NADPH diaphorase staining are co-localised in neurons closely associated with the vasculature in rat and human retina. Brain Res 684:36–46

Ruskell GL (1965) The orbital distribution of the sphenopalatine ganglion in the rabbit. In: Rohen JW (ed) The structure of the eye. II. Symposium. Schattauer, Stuttgart, pp 355–368

Ruskell GL (1970a) The orbital branches of the pterygopalatine ganglion and their relationship with internal carotid nerve branches in primates. J Anat 106:323–339

Ruskell GL (1970b) An ocular parasympathetic nerve pathway of facial nerve origin and its influence on intraocular pressure. Exp Eye Res 10:319–330

Ruskell GL (1971) Facial parasympathetic innervation of the choroidal blood-vessels in monkeys. Exp Eye Res 12:166–172

Ruskell GL, Griffiths T (1979) Peripheral nerve pathway to the ciliary muscle. Exp Eye Res 28:277–284

Samuel U, Lütjen-Drecoll E, Tamm ER (1996) Gap junctions are found between iris sphincter smooth muscle cells but not in the ciliary muscle of human and monkey eyes. Exp Eye Res 63:187–192

Sanders KM, Ward SM (1992) Nitric oxide as a mediator of nonadrenergic noncholinergic neurotransmission. Am J Physiol 262:G379-G392

Schmidt HHHW, Walter U (1994) NO at work. Cell 78:919–925

Schuman JS, Erickson K, Nathanson JA (1994) Nitrovasodilator effects on intraocular pressure and outflow facility in monkeys. Exp Eye Res 58:99–105

Sienkiewicz W, Kaleczyc J, Majewski M, Lakomy M (1995) NADPH-diaphorase-containing cerebrovascular nerve fibres and their possible origin in the pig. J Brain Res 36:353–363

Stjernschantz J, Bill A (1979) Effect of intracranial stimulation of the oculomotor nerve on ocular blood flow in the monkey, cat and rabbit. Invest Ophthalmol Vis Sci 18:99–103

Stjernschantz J, Bill A (1980) Vasomotor effects of facial nerve stimulation: noncholinergic vasodilation in the eye. Acta Physiol Scand 109:45–50

Stöhr P (1957) Handbuch der mikroskopischen Anatomie. V. Mikroskopische Anatomie des vegetativen Nervensystems. Springer, Berlin Heidelberg

Stone RA, Kuwayama Y, Laties AM (1987) Regulatory peptides in the eye. Experientia 43:791–800

Su E-N, Alder VA, Yu D-Y, Cringle SJ (1994) Adrenergic and nitrergic neurotransmitters are released by the autonomic system of the pig long posterior ciliary artery. Curr Eye Res 13:907–917

Sun W, Erichsen JT, May PJ (1994) NADPH-diaphorase reactivity in ciliary ganglion neurons: a comparison of distributions in the pigeon, cat, and monkey. Vis Neurosci 11:1027–1031

Suzuki N, Fukuuchi Y, Koto A, Naganuma Y, Isozumi K, Matsuoka S, Gotoh J, Shimizu T (1993) Cerebrovascular NADPH diaphorase-containing nerve fibers in the rat. Neurosci Lett 151:1–3

Szmydynger-Chodobska J, Monfils PR, Lin AY-J, Rahman MP, Johanson CE, Chodobski A (1996) NADPH-diaphorase histochemistry of rat choroid plexus blood vessels and epithelium. Neurosci Lett 208:179–182

Talmage EK, Mawe GM (1993) NADPH-diaphorase and VIP are colocalized in neurons of gallbladder ganglia. J Auton Nerv Syst 43:83–90

Tamm E, Flügel C, Baur A, Lütjen-Drecoll E (1991a) Cell cultures of human ciliary muscle: growth, ultrastructural and immunocytochemical characteristics. Exp Eye Res 53:375–387

Tamm E, Lütjen-Drecoll E, Jungkunz W, Rohen JW (1991b) Posterior attachment of ciliary muscle in young, accommodating old, presbyopic monkeys. Invest Ophthalmol Vis Sci 32:1678–1692

Tamm E, Flügel C, Stefani FH, Rohen JW (1992) Contractile cells in the human scleral spur. Exp Eye Res 54:531–543

Tamm ER, Lütjen-Drecoll E (1997) Nitrergic nerve cells in the primate ciliary muscle are only present in species with a fovea centralis. Ophthalmologica 211:201–204

Tamm ER, Lütjen-Drecoll E (1996a) Functional morphology and origin of nitrergic nerves in the human eye. Exp Eye Res 63(Suppl):S.151

Tamm ER, Lütjen-Drecoll E (1996b) Ciliary body. Microsc Res Tech 33:390–439

Tamm ER, Flügel-Koch C, Mayer B, Lütjen-Drecoll E (1995a) Nerve cells in the human ciliary muscle: ultrastructural and immunocytochemical characterization. Invest Ophthalmol Vis Sci 36:414–426

Tamm ER, Koch TA, Mayer B, Stefani FH, Lütjen-Drecoll E (1995b) Innervation of myofibroblast-like scleral spur cells in human and monkey eyes. Invest Ophthalmol Vis Sci 36:1633–1644

Tiffany JM (1995) Physiological functions of the meibomian glands. In: Osborne NN, Chader G (eds) Progress in retinal and eye research. Elsevier Science, Amsterdam, pp 47–74

Toda N (1995) Nitroxidergic nerves and hypertension. Hypertens Res 18:19–26

Toda N, Ayajiki K, Yoshida K, Kimura H, Okamura T (1993) Impairment by damage of the pterygopalatine ganglion of nitroxidergic vasodilator nerve function in canine cerebral and retinal arteries. Circ Res 72:206–213

Toda N, Kitamura Y, Okamura T (1994) Role of nitroxidergic nerve in dog retinal arterioles in vivo and arteries in vitro. Am J Pathol 266:H1985–H1992

Toda N, Toda M, Ayajiki K, Okamura T (1996) Monkey central retinal artery is innervated by nitroxidergic vasodilator nerves. Invest Ophthalmol Vis Sci 37:2177–2184

Toris CB, Pederson JE (1987) Aqueous humor dynamics in experimental iridocyclitis. Invest Ophthalmol Vis Sci 28:477–481

Törnquist P, Alm A (1979) Retinal and choroidal contribution to retinal metabolism in vivo: a study in pigs. Acta Physiol Scand 106:351–357

Törnqvist G (1966) Effect of cervical sympathetic stimulation on accommodation in monkeys: an example of a beta-adrenergic, inhibitory effect. Acta Physiol Scand 67:363–372

Törnqvist G (1967a) The relative importance of the parasympathetic and sympathetic nervous system for accommodation in monkeys. Invest Ophthalmol Vis Sci 6:612–617

Törnqvist G (1967b) Accommodation in monkeys: some pharmacological and physiological aspects. Acta Ophthalmol (Copenh) 45:1–32

Tracey WR, Nakane M, Pollock JS, Förstermann U (1993) Nitric oxide synthases in neuronal cells, macrophages and endothelium are NADPH diaphorases, but represent only a fraction of total cellular NADPH diaphorase activity. Biochem Biophys Res Commun 195:1035–1040

Tripathi RC, Tripathi BJ (1984) Anatomy of the human eye, orbit, and adnexa. In: Davson H (ed) The eye. Vol 1a: Vegetative physiology and biochemistry. Academic, San Diego, pp 1–268

Uddman R, Alumets J, Ehinger B, Hakanson R, Loren I, Sundler F (1980) Vasoactive intestinal peptide nerves in ocular and orbital structures of the cat. Invest Ophthalmol Vis Sci 19:878–885

Ueno M, Naumann GOH (1989) Uveal damage in secondary glaucoma. Graefes Arch Clin Exp Ophthalmol 227:380–383

Unger WG (1989) Mediation of the ocular response to injury and irritation: peptides versus prostaglandins. In: Bito LZ, Stjernschantz J (eds) The ocular effects of prostaglandins and other eicosanoids: proceedings in clinical and biological research. Vol 312. Liss, New York, pp 293–328

Van Alphen GWHM, Robinette SL, Macri FJ (1962) Drug effects on ciliary muscle and choroid preparations in vitro. Arch Ophthalmol 68:81–93

Van der Werf F (1993) Innervation of the lacrimal gland in the cynomolgus monkey: a retrograde tracing and immunohistochemical study. In: van der Werf F (ed) Autonomic and sensory innervation of some orbital structures in the primate. Thesis, Universiteit van Amsterdam, Amsterdam, pp 51–70

Van der Zypen E (1967) Licht- und elektronenmikroskopische Untersuchungen über den Bau und die Innervation des Ziliarmuskels bei Mensch und Affe (*Cercopithecus ethiops*). Graefes Arch Klin Exp Ophthalmol 174:143–168

Wang I, Kondo M, Bill A (1995) Vascular responses to flickering light in the retina in cats and monkeys: effect of L-NAME. Acta Physiol Scand 153:39A

Wang ZY, Alm P, Hakanson R (1995) Distribution and effects of pituitary adenylate cyclase activating polypeptide in the rabbit eye. Neuroscience 69:297–308

Warwick R (1954) The ocular parasympathetic nerve supply and its mesencephalic sources. J Anat 88:195–203

Wiedenmann B, Franke WW (1985) Identification and localization of synaptophysin, an integral membrane glycoprotein of Mr 38000 characteristic of presynaptic vesicles. Cell 41:1017–1028

Wiederholt M, Sturm A, Lepple-Wienhues A (1994) Relaxation of trabecular meshwork and ciliary muscle by release of nitric oxide. Invest Ophthalmol Vis Sci 35:2515–2520

Wiederholt M, Bielka S, Schweig F, Lütjen-Drecoll E, Lepple-Wienhues A (1996) Regulation of outflow rate and resistance in the perfused anterior segment of the bovine eye. Exp Eye Res 61:223–234

Wiencke AK, Nilsson H, Nielsen PJ, Nyborg NCB (1994) Nonadrenergic noncholinergic vasodilation in bovine ciliary artery involves CGRP and neurogenic nitric oxide. Invest Ophthalmol Vis Sci 35:3268–3277

Wilcox LM, Keough EM, Connolly RJ, Hote CE (1980) The contribution of blood flow by the anterior ciliary arteries to the anterior segment in the primate eye. Exp Eye Res 30:167–174

Wilkinson KD, Lee K, Deshapande S, Duerksen-Hughes P, Boss JM, Pohl J (1989) The neuron-specific protein PGP 9.5 is a ubiquitin carboxyl-terminal hydrolase. Science 246:670–673

Wizemann A, Wizemann V (1980) Untersuchungen zur ambulanten und perioperativen Augendrucksenkung mit organischen Nitraten. Klin Monatsbl Augenheilkd 177:292–295

Yamamoto R, Bredt DS, Snyder SH, Stone RA (1993) The localization of nitric oxide synthase in the rat eye and related cranial ganglia. Neuroscience 54:189–200

Ye X, Laties AM, Stone RA (1990) Peptidergic innervation of the retinal vasculature and the optic nerve head. Invest Ophthalmol Vis Sci 31:1731–1737

Yoshida K, Okamura T, Kimura H, Bredt DS, Snyder SH, Toda N (1993) Nitric oxide synthase-immunoreactive nerve fibers in dog cerebral and peripheral arteries. Brain Res 629:67–72

Yoshida K, Okamura T, Toda N (1994) Histological and functional studies on the nitroxidergic nerve innervating monkey cerebral, mesenteric and temporal arteries. Jpn J Pharmacol 65:351–359

Zagvazdin YS, Fitzgerald MEC, Sancesario G, Reiner A (1996) Neural nitric oxide mediates Edinger-Westphal nucleus evoked increase in choroidal blood flow in the pigeon. Invest Ophthalmol Vis Sci 37:666–672

Zhang YL, Tan CK, Wong WC (1994a) Localisation of substance P-like immunoreactivity in the ciliary ganglia of monkey (*Macaca fascicularis*) and cat: a light- and electron-microscopic study. Cell Tissue Res 276:163–171

Zhang YL, Tan CK, Wong WC (1994b) The ciliary ganglion of the monkey: a light and electron microscope study. J Anat 184:251–260

III. Ocular Bloodflow: Ocular Autonomic Neurotransmission

Cell Culture Studies of Oxygen, Nitric Oxide, and Retinal Pericytes' Contractile Tone

Ivan O. Haefliger[1] and Douglas R. Anderson[2]

Introduction

Risk Factors Associated with Glaucoma

Although the pathophysiology of glaucoma is still a matter of debate, it is clear that several conditions represent risk factors for this optic nerve head neuropathy (Shields 1992). Classically, the damage found with glaucoma has been linked to high introcular pressure (IOP) (Anderson 1989). In support of this linkage is that subjects develop unilateral glaucoma after a unilateral posttraumatic increase of IOP. In these patients it is only the eye with a high IOP that has glaucomatous damage, not the eye with normal pressure. Furthermore, the progression of damage can be stopped by reducing the IOP to normal values, which demonstrates that the high IOP, rather than the trauma itself, can lead to glaucoma (Anderson 1989; Shields 1992; Haefliger and Flammer 1997a). The importance of the role of IOP in glaucoma has been further supported by experimental animal models, in which glaucomatous cupping could be elicited after several months of an artificial increase in IOP (Shields 1992). These examples, as well as the common observation that most patients with primary open-angle glaucoma also have some elevation of IOP, have led us to assume that glaucoma was due only to an increase of the IOP and, by definition, that glaucoma is intolerably high IOP.

Although it is true that subjects who have glaucoma damage tend to have high IOP, the link is much weaker if one looks at this situation from a different point of view. For example, consider the course of a population of subjects who all have the same high IOP (e.g., 26 mmHg) and thus are exposed to the same IOP risk factor. Some of these subjects tolerate this pressure for years without apparent damage, whereas others develop glaucoma within a few months. This obser-

[1] Laboratory of Ocular Pharmacology and Physiology, University Eye Clinic Basel, Mittlere Strasse 91, CH-4051 Basel, Switzerland
[2] Department of Ophthalmology, Bascom Palmer Eye Institute, University of Miami School of Medicine, 900 NW 17th, Miami, FL 33136, USA

vation clearly demonstrates that in addition to the IOP there must exist other factors that make some subjects more sensitive than the others to the IOP (Anderson 1989; Flammer et al. 1992; Flammer 1996; Spaeth 1996). One can even find some conditions in which these additional risk factors seem so predominant that subjects can develop glaucoma despite the fact that their IOP is normal. These patients have normal-tension glaucoma (NTG) (Shields 1992).

What could be the nature of these additional risk factors? Epidemiological studies have pointed out that, in addition to the high IOP, certain conditions were found more often in glaucoma patients. Risk factors may include a family history of visual loss from glaucomatous scotoma, gender (women seem to be more affected by NTG than men), age, and the presence of myopia. As can be seen, most of these conditions reflect a genetic predisposition for which, unfortunately, there is no therapy as yet (Flammer 1996; Haefliger and Flammer 1997b).

Vascular Risk Factors and Glaucoma

Studies have shown that some glaucoma patients, especially those who have rapid progression of damage, had a marked tendency for systemic hypotension (Demailly et al. 1984; Kaiser and Flammer 1991; Kaiser et al. 1993; Béchetoille and Bresson-Dumont 1994; Hayreh et al. 1994; Graham et al. 1995) and peripheral vasospasm (Guthauser et al. 1988; Gasser and Flammer 1991). These entities are vascular risk factors for which there is potential therapy (Flammer 1996; Haefliger and Flammer 1997a).

Strikingly, the visual field of some vasospastic glaucoma patients can be modulated by vasoactive stimuli such as the inhalation of carbogen, a mixture of oxygen and carbon dioxide, which is known to induce vasodilation. Breathing carbogen transiently improves their visual field (Pillunat et al. 1994; Pillunat 1997). Such improvement remains if these patients are treated for several months with calcium channel blockers (Kitazawa et al. 1989; Gasser and Flammer 1990; Flammer 1993, 1996; Pillunat et al. 1994; Kanellopoulos et al. 1996).

It appears that glaucoma is an optic nerve head (ONH) neuropathy, and at least in some patients with glaucoma dysregulation of ocular blood flow might be important in the pathogenesis of this neuropathy (Anderson 1996; Flammer 1996). An obvious link exists between the IOP and the circulation in the ONH. Specifically, an elevated IOP increases the venous pressure by compressing veins at their exit point from the eye and consequently challenges the circulation in the eye, particularly in the ONH (Anderson and Quigley 1992; Haefliger and Anderson 1996a). However, because there is autoregulation, the blood flow is maintained in the ONH and the retina (Geijer and Bill 1979; Weinstein et al. 1982, 1983; Sossi and Anderson 1983; Pillunat et al. 1997). There is evidence of deficient autoregulation in patients with glaucoma (Sinclair et al. 1982; Grunwald et al. 1984; Petrig et al. 1985; Pillunat et al. 1985; Riva et al. 1986; Robinson et al. 1986; Ulrich et al. 1986; Pillunat et al. 1987; Robert et al. 1989). Because dysregulation in the ONH ocular blood flow might be important in the pathogenesis of glaucoma we became interested in the circulation of the ONH.

Vascular Diameter and Blood Flow

The circulation within the ONH is composed of a dense capillary network. On the vascular casting of a monkey ONH, represented on Fig. 1, one can appreciate the density of the capillary network that interconnects in a continuous manner with the capillaries of the retinal circulation (Anderson 1970; Anderson and Braverman 1976). The retinal capillary network (Toussaint et al. 1961; Kuwabara and Cogan 1963) consists of vessels distributed in parallel and in series. The global hemodynamic of this vascular structure is characterized by a small arteriovenous pressure difference and a small overall vascular resistance because of numerous parallel channels, despite the fact that, taken individually, the resistance of a single capillary can be high by reason of its small diameter.

In vessels larger than 200 μm diameter, the flow is governed by Poiseuille's law (Guyton 1991), which states that the flow (Q) through a vessel is proportional to the perfusion pressure [the arteriovenous pressure difference $(P_a - P_v)$]; proportional to the fourth power of the radius (r); and inversely proportional to the length (l) of the vessel and the viscosity (η) of the fluid (two factors that can be considered constant in a given vessel larger than an arteriole).

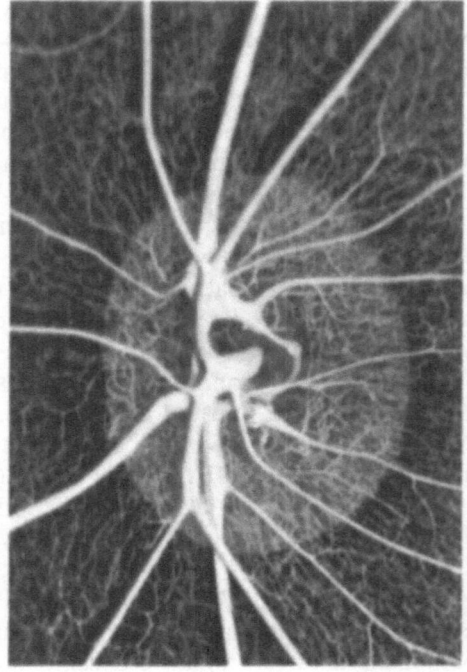

Fig. 1. Vascular casting of a monkey optic nerve head demonstrating the presence of a dense capillary network in continuity with the capillary network of the retinal circulation. (From Anderson 1970, with permission)

$$Q = \pi(P_a - P_v)r^4/8\eta l$$

as: $\pi/8\eta l = constant$

$$Q = (P_a - P_v)r^4$$

This theoretical equation reflects more or less what one would intuitively expect, in the sense that the blood flow through a vessel increases if the arterial pressure is increased or the venous pressure decreased, or if the radius of the vessel becomes larger. This equation thus means that a reduction in the diameter of a vessel in the range of 10% or 25% can account for a decrease in flow through the vessel of 30% and 70%, respectively. Such an example clearly illustrates the powerful influence of small changes in the diameter of a vessel on the blood flow through this vessel (Haefliger 1995).

In theory, the Poiseuille's law does not apply to the vessels of the microcirculation. When in vivo experiments were conducted in the cat mesenteric microcirculation (Lipowsky et al. 1978), it was empirically found that, as in larger vessels, the resistance to flow was in fact virtually proportional to the fourth power of the vascular diameter (in precapillary arterioles to postcapillary venules) and that the highest resistance was found in capillaries (Fig. 2). In other words, it appears that in both large and small vessels, even those of the size of a capillary, a relation to the power of four exists between blood flow (or resistance) and the vascular diameter (Zweifach and Lipowsky 1984). Therefore if small vessels of the size of a capillary had the capacity to change their diameters, they could modulate blood flow passing through their lumen.

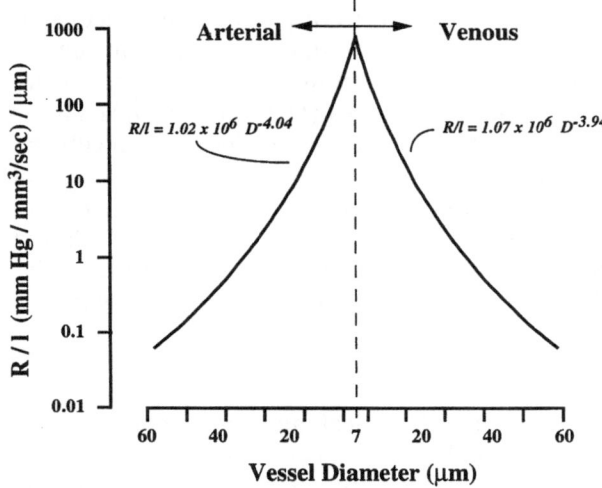

FIG. 2. Measurement of the arteriovenous distribution of intravascular resistance per unit length in single unbranched vessels of the cat's mesenteric microcirculation. A relation close to the power of four could be observed between the vascular diameter and the resistance. R/l, resistance (R) per unit of length (l); D, vascular diameter. (Modified from Lipowsky et al. 1978, with permission)

FIG. 3. Simplified drawing of a capillary shows an internal lining of
endothelium surrounded by a layer
of pericytes

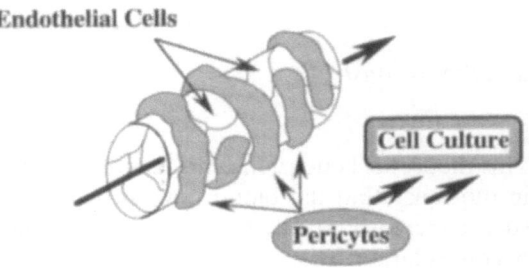

Endothelial Cells

Cell Culture

Pericytes

The capillaries of the ONH and the retina, like the capillaries of the cerebral circulation, are composed of a continuous internal layer of endothelial
cells linked by tight junctions, a layer of endothelial cells surrounded by an important layer of pericytes (Frank et al. 1987, 1990) (Fig. 3). Like vascular smooth
muscle cells, pericytes are contractile (Tilton et al. 1979; Joyce et al. 1984, 1985a,
1985b; Chan et al. 1986; Kelley et al. 1987, 1988; Das et al. 1988; Lee et al. 1989;
Dodge et al. 1991; Chakravarthy et al. 1992; Ferrari-Dileo et al. 1992; Helbig et
al. 1992). Pericytes respond to endothelium-derived vasoactive substances, such
as the contracting factors thromboxane A_2 (Dodge et al. 1991) and endothelin-
1 (Chakravarthy et al. 1992), or the relaxing factor prostacyclin (Dodge et al.
1991), known to be released in the ophthalmic circulation (Benedito et al. 1991a,
1991b; Hoste and Andries 1991; Haefliger et al. 1992, 1993a, 1993b, 1994a, 1994b).

Material and Methods

Pericyte Isolation and Cell Culture

As it is difficult to study the contractile properties of pericytes in intact capillaries (in vivo or in vitro), we have chosen to study them in cell culture.
We have taken bovine eyes and grown pericytes from the retinal microcirculation by dissecting free retinas of 30–40 bovine eyes (D'Amore 1990; Haefliger
et al. 1994b; Anderson and Davis 1996). Retinas were then minced and incubated
for 1h in phosphate-buffered saline (PBS) containing 0.2% collagenase and
0.2% bovine serum albumin. After filtration through a Nitex mesh, the cells were
retrieved and rinsed by serial centrifugation ($800g$), resuspended, and cultured
in 75-cm square flasks in Dulbecco's modified eagle medium (DMEM) containing 10% fetal bovine serum, fungizone 50mg/ml, and gentamicin 1.25μg/ml.

Pericyte Identification

The cells cultured had the typical morphological appearance of pericytes
(D'Amore 1990; Anderson and Davis 1996) and stained positively for the antiganglioside antibody 3G5, which marks pericytes but not vascular smooth muscle

cells (Nayak et al. 1988). These cultured pericytes did not express the glial fibrillar acidic protein found in cultured astrocytes (Aotaki-Keen et al. 1991); they lacked the retinal pigment epithelium's ability to phagocytize rod outer segments (McLaren et al. 1993), and they also lacked the ability to take up fluorescein-labeled acetylated low-density lipoprotein (LDL), a characteristic feature of endothelial cells (Voyata et al. 1984), which are the main potential contaminants of pericyte cultures obtained from retina. From these general features in cultured pericytes (Anderson and Davis 1996), we ensured the identification of cells and the absence of endothelium for each lot of cells used in our experiments by documenting binding of the antiganglioside 3G5 and the absence of cells that take up acetylated LDL, respectively (Fig. 5).

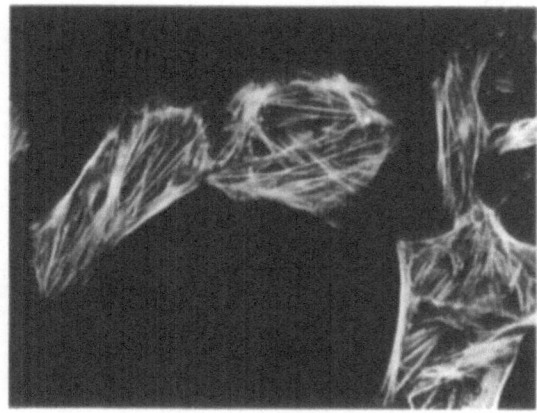

FIG. 4. Bovine retinal pericytes in culture, demonstrating myosin by indirect immunofluorescence microscopy. (×320) (From Anderson and Davis 1996, with permission)

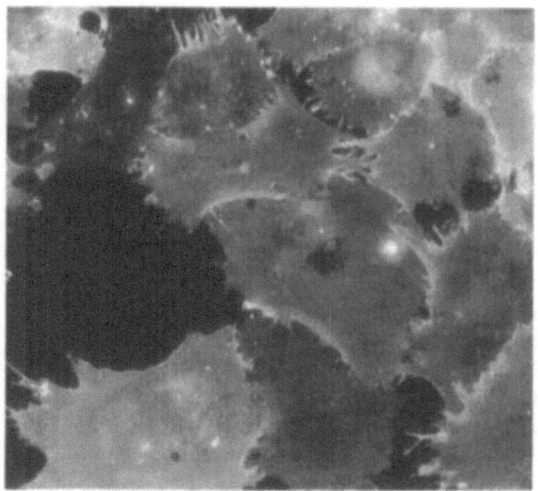

FIG. 5. Reaction of bovine retinal pericytes in culture with antiganglioside monoclonal antibody 3G5. (×320) (From Anderson and Davis 1996, with permission)

Culture of Pericytes on Silicone Membrane

First and second passages of bovine retinal pericytes were grown in DMEM for 3–7 days on a thin silicone membrane coating the bottom of a petri dish (dimethylpolysiloxane, 60000 centipoise viscosity silicone fluid surface hardened by flame). When these contractile cells attach to the silicone membrane and grow on its surface, they place the underlying elastic silicone membrane under tension, which results in a series of small wrinkles that can be visualized with a phase-contrast inverted microscope (Zeiss, Oberkochen, Germany) (Harris et al. 1980; Haefliger et al. 1994b). Pericytes in these cultures spontaneously exhibited a contractile tone that produced a wrinkled silicone surface underneath them (Haefliger et al. 1994b). To enhance the reproducibility of this method, every experiment was matched with a control conducted with the same batch of cells and same batch of silicone membranes.

Observation of Pericytes on Silicone Membrane

To observe the cells and the wrinkles the whole petri dish was placed on the stage of a microscope enclosed in a transparent chamber that maintained the cells at a constant 37°C temperature while allowing multiple fluid exchanges by a suction-perfusion system (Fig. 6). Within the chamber the room-air atmosphere

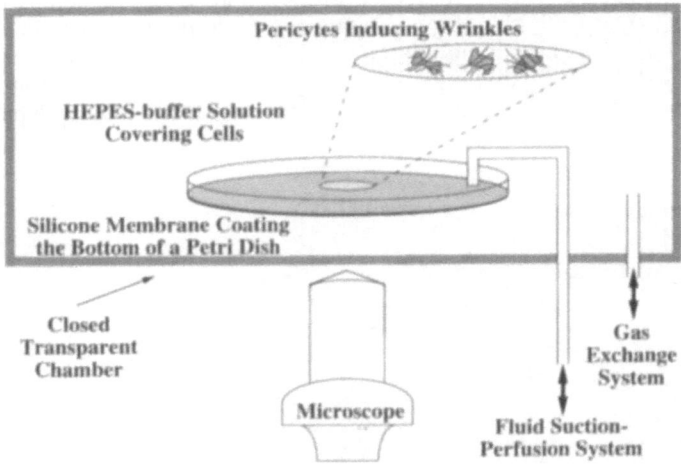

FIG. 6. Setup used to study changes in pericytes' contractile tone. Pericytes were grown on a silicone membrane coating the bottom of a petri dish. Cells were observed by a phase-contrast inverted microscope. By a suction-perfusion system, the HEPES buffer solution covering the cells could be exchanged and retinal pericytes exposed to various drugs. By a gas exchange system, the atmosphere in the closed transparent chamber containing the petri dish could be saturated with different concentrations of oxygen or nitrogen (Haefliger and Anderson 1997a, 1997b)

could be saturated by insulfating oxygen or nitrogen into the chamber under positive pressure. The microscope was connected to a video camera system (Panasonic, Osaka, Japan) and a printer (Mitsubishi, Cypress, CA, USA), and the number of wrinkles appearing on the prints was tabulated with a Zeiss Videoplan 2 Image Analysis System (Kontron, Eching, Germany) (Haefliger et al. 1994b). Fields of view were selected in which the cells were sufficiently separated, but not in a cluster, so the wrinkles due to individual cells could be easily analyzed (1–5 cells per field of view).

Experimental Conditions of the Study

Once the petri dish had been placed in the chamber, the DMEM solution was immediately replaced with a HEPES (pH 7.4)-buffered solution (NaCl 140 mM, KCl 5 mM, CaCl$_2$ 1 mM, MgCl$_2$ 1.5 mM, HEPES 5 mM, HEPES sodium salt 5 mM, glucose 10 mM), a solution that was then used in all subsequent experimental manipulations. After 20 min of adaptation in this solution, the pericytes, which tend to have a high level of basal contractile tone, were exposed to sodium nitroprusside (SNP), 8-bromo-cyclic guanosine monophosphate (cGMP), 3-morpholino-sydnonimine (SIN-1), the atrial natriuretic peptide (ANP), or forskolin in the presence or absence of methylene blue or hemoglobin or under different conditions of oxygenation. In all cases, control runs alternated with experimental runs.

Hemoglobin Preparation

Bovine hemoglobin (crystallized, dialyzed, and lyophilized) was placed in PBS (pH 7.4), converted to oxyhemoglobin with sodium hydrosulfite, redialyzed, and deoxygenated by bubbling with 100% nitrogen. Small aliquots were stored at −80°C until thawed for use. Low concentrations of methylhemoglobin were ensured by spectrophotometric assay at the time of experimentation.

Changes in Oxygen Concentration

Hypoxic or hyperoxic conditions were obtained by bubbling the HEPES-buffered solution for 10 min with 100% nitrogen or 100% oxygen prior to its infusion in the petri dish. Once in the petri dish the atmosphere surrounding the solution was immediately saturated with either 100% nitrogen or 100% oxygen. In preliminary experiments it was shown that after bubbling the petri dish solution for 10 min with 100% nitrogen or 100% oxygen the partial pressures of oxygen in the solution were 2% and 97%, respectively, conditions that are clearly hypoxic and hyperoxic (Haefliger and Anderson 1997a, 1997b). We studied specifically, the effect of hypoxic, normoxic, and hyperoxic conditions regarding the basal resting tone of pericytes, and the relaxation induced by sodium nitroprusside, SIN-1, ANP, or forskolin.

Assessment of the Data and Statistical Analysis

Changes in pericyte contractile tone were quantified by counting the changes in the number of the wrinkles (Haefliger et al. 1994b). Measurements were usually performed at 2, 5, and 10 min. With all experimental protocols the maximal effect was maintained until the final measurement at 10 min; therefore measurements at 10 min were compared. Results are expressed as the percent of the number wrinkles lost, reflecting the pericyte basal tone. The data are given as the means and the standard errors of the mean (mean ± SEM), and the results are compared using an unpaired Student's t-test, with a $P < 0.05$ considered significant.

Drugs Used During the Study

Atrial natriuretic peptide, sodium nitroprusside, forskolin, fungizone, collagenase, HEPES, dimethylpolysiloxane, and bovine serum albumin were obtained from Sigma (St. Louis, MO, USA). SIN-1, the active metabolite of molsidomine, was a gift from Hoechst Pharmaceuticals (Paris, France). Gentamicin and the fetal bovine serum were purchased from Gibco (Gaithersburg, MD, USA); and the oxygen and the nitrogen were from Liquid Carbonics (Miami, FL, USA). Sodium nitroprusside was dissolved in distilled water, and a fresh solution was made daily. Forskolin was dissolved in alcohol and stored in stock solution (–20°C).

Results

Retinal Pericytes: Contractile Cells

The original experiments measured the changes in pericytes and their wrinkles before and after a 10-min exposure to two concentrations (3×10^{-7}M and 3×10^{-5}M) of sodium nitroprusside. The loss of wrinkles evoked by this drug is clear (Fig. 7).

Pericyte Relaxation Induced by Sodium Nitroprusside

Even though the changes in the contractile tone of pericytes are not directly measured in millinewtons, counting the number of wrinkles provides reproducible, reliable results. By exposing pericytes to increasing concentrations of sodium nitroprusside it was possible to construct a concentration–response curve and to calculate the ER_{50} (50% evoked response) of this drug, in this case 1 μM (Figs. 8, 9). This example clearly shows that pericytes from the retinal microcirculation are contractile cells and that they can be relaxed by a drug such as sodium nitroprusside.

To demonstrate the mechanism by which sodium nitroprusside relaxes pericytes, experiments were conducted with 3 μm sodium nitroprusside, which evokes

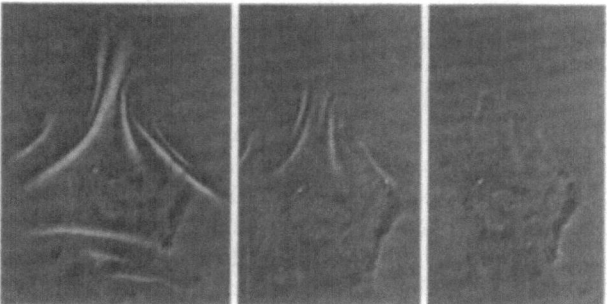

FIG. 7. Wrinkles induced by a single pericyte grown on a thin sheet of silicone and observed with a phase contrast-inverted microscope (*left panel*). Note the loss of wrinkles induced by 10 min of exposure to 3 and 100 μM sodium nitroprusside, respectively (*middle* and *right panels*). (×1200) (From Haefliger et al. 1994b, with permission)

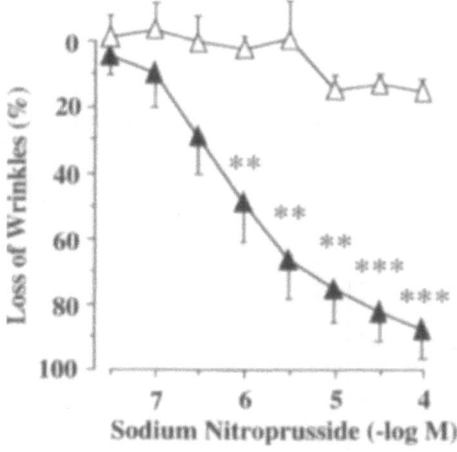

FIG. 8. Concentration–response curve of the relaxing effect of sodium nitroprusside (*SNP*) in comparison to time-control experiments. In a concentration dependent-manner, sodium nitroprusside evoked pericytes' relaxation and thus a loss of wrinkles. *Open triangles*, controls; *filled triangles*, SNP. **$p < 0.01$; ***$p < 0.001$. Each triangle represents five tests. (From Haefliger et al. 1994b, with permission)

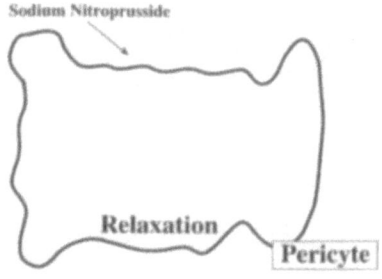

FIG. 9. Simplified drawing illustrates the fact that retinal pericytes are contractile cells that can be relaxed by sodium nitroprusside

somewhat more than 50% relaxation. Hence the relaxing effect of this drug was modulated by different substances.

Effect of NO

In the presence of 10μM hemoglobin (Hb), a scavenger of NO, the relaxation induced by sodium nitroprusside was significantly inhibited, demonstrating that sodium nitroprusside relaxes pericytes most likely through release of NO (Figs. 10, 11) (Haefliger and Anderson 1996b). In contrast, relaxation induced by other substances, such as ANP in particular, was not abolished by Hb. Thus Hb does not interfere with the relaxation properties of pericytes mediated by cGMP, except when NO is the stimulant of guanylate cyclase (see Figs. 26 and 27 below) (Haefliger and Anderson, 1997b).

Guanylate Cyclase Activation

It is known that in smooth muscle cells NO can stimulate guanylate cyclase, which is responsible for production of the second messenger cGMP. To address

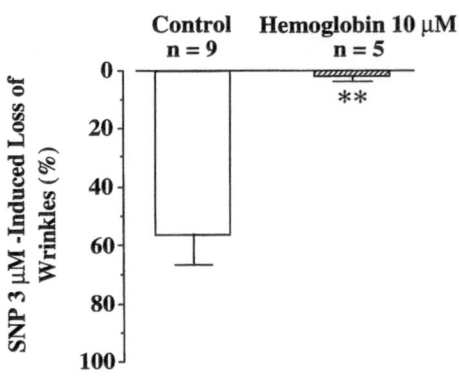

Fig. 10. Inhibitory effect of hemoglobin 10μM, a scavenger of nitric oxide (*NO*), on the relaxation of retinal pericytes evoked by sodium nitroprusside 3μM. **$p < 0.01$. (Modified from Haefliger and Anderson 1996b, with permission)

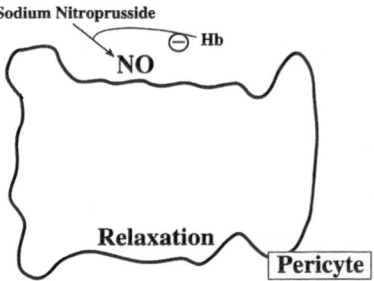

Fig. 11. Simplified drawing illustrates the fact that NO mediates the sodium nitroprusside-induced relaxation of retinal pericytes. *Hb*, hemoglobin

whether NO activates guanylate cyclase in pericytes, cells were exposed to 3 μM sodium nitroprusside in the absence or presence of increasing concentrations of an inhibitor of the enzyme guanylate cyclase, methylene blue (Figs. 12, 13). The half-maximal relaxation produced by 3 μM sodium nitroprusside was progressively inhibited by increasing concentrations of methylene blue (0.03, 0.01, 0.3 μM).

To demonstrate that in pericytes the relaxation evoked by NO is indeed associated with an increase of the second messenger cGMP, experiments were conducted in which the cGMP production was measured by radioimmunoassay. Exposure of pericytes to increasing concentrations of sodium nitroprusside evoked a huge increase in cGMP production: 14- and 44-fold increases for concentrations of 3 and 100 μM, respectively (Figs. 14, 15).

Furthermore, the increased cGMP production induced by 3 μM of sodium nitroprusside was significantly inhibited by 0.3 μM methylene blue (Figs. 16, 17).

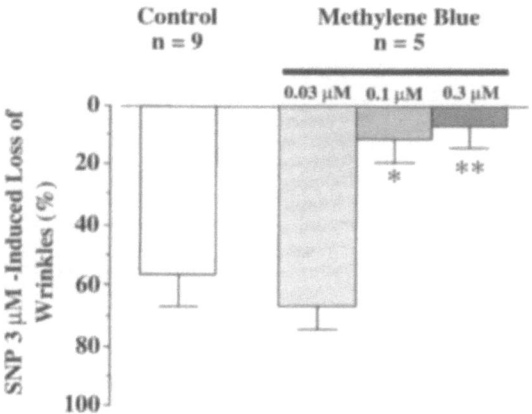

Fig. 12. Inhibitory effect of increasing concentrations of the inhibitor of guanylate cyclase, methylene blue, on the relaxation of retinal pericytes evoked by sodium nitroprusside 3 μM. *$p < 0.05$; **$p < 0.01$. (Modified from Haefliger et al. 1994b, with permission)

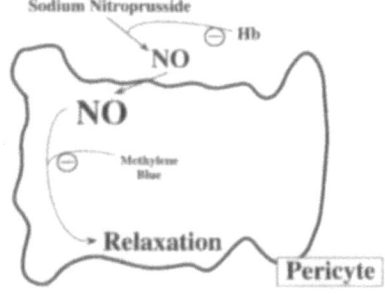

Fig. 13. Simplified drawing illustrates the fact that methylene blue inhibits NO-induced relaxation in retinal pericytes

FIG. 14. Effect of increasing concentrations of sodium nitroprusside on the production of cyclic guanosine monophosphate (cGMP) by retinal pericytes. (From Haefliger et al. 1994b, with permission)

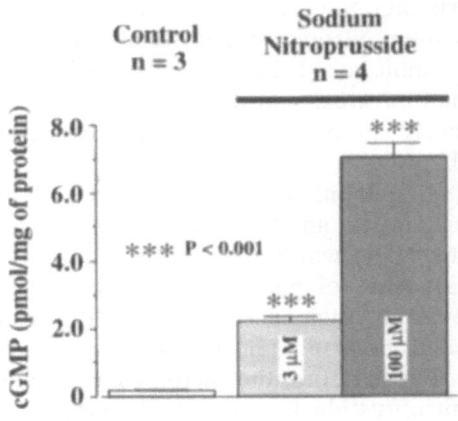

FIG. 15. Simplified drawing illustrates that NO increases cGMP production in retinal pericytes. *GTP*, guanosine triphosphate

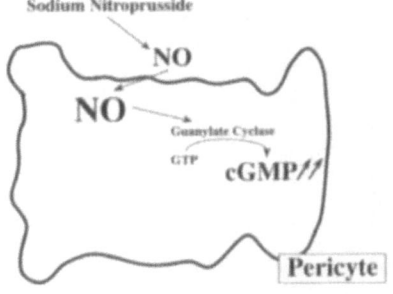

FIG. 16. Inhibitory effect of the inhibitor of guanylate cyclase, methylene blue, on the 3 μM sodium nitroprusside-induced cGMP production. ***$p < 0.001$. (From Haefliger et al. 1994b, with permission)

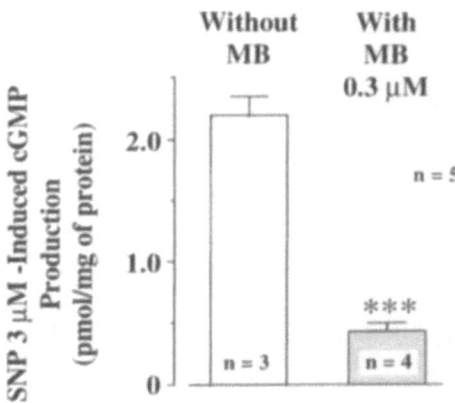

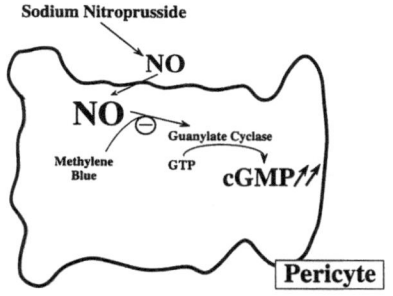

FIG. 17. Simplified drawing illustrates that methylene blue inhibits NO-induced cGMP production in retinal pericytes

This experiment demonstrated that the NO-induced relaxation of pericytes is associated with an increase in cGMP, and that the relaxation and increased cGMP production are inhibited by methylene blue.

cGMP-Induced Relaxation: Effect of Methylene Blue

To exclude that methylene blue could interfere with the relaxing properties of pericytes by means other than guanylate cyclase inhibition and to confirm further that cGMP is indeed able to relax these cells, pericytes were exposed to a cGMP stable analog, 8-bromoguanosine 3′:5′-monophosphate (8-bromo-cGMP), which penetrates the plasma membrane. As expected, after exposure to increasing concentrations of 8-bromo-cGMP, pericytes relaxed in a concentration-dependent manner (Figs. 18, 19). Furthermore, the response to 30 μM 8-bromo-cGMP was not inhibited by 0.3 μM methylene blue, although at this methylene blue concentration both the relaxation and the production of cGMP evoked by 3 μM sodium nitroprusside were strongly inhibited.

In summary, these results demonstrate that pericytes are contractile cells; that the relaxation to NO involves the production of cGMP, which mediates the relaxation; and that methylene blue does not interact with the relaxing properties of cGMP but, rather, affects its production (Fig. 20).

NO-Induced Relaxation: Effect of Oxygen

Under physiological conditions, circulation in the optic nerve head and retina is autoregulated, particularly in response to oxygen. Indeed, hypoxia results in increased blood flow, and hyperoxia leads to decreased flow (Riva et al. 1983; Pournaras 1996). In addition, in the ophthalmic circulation there is a constant endothelium release of NO, maintaining this vascular bed in a constant state of mid-dilation (Haefliger et al. 1993b, 1994a; Donati et al. 1995). As NO is known to be degraded by oxygen (Wink et al. 1996), the question of possible oxygen interference with the relaxation of NO derived from SNP or SIN-1 became relevant.

FIG. 18. Relaxing effect of 8-bromo-cGMP on the contractile tone of bovine retinal pericytes and the lack of inhibitory effect of methylene blue (*MB*). *$P < 0.05$; **$P < 0.01$. (Modified from Haefliger et al. 1994b, with permission)

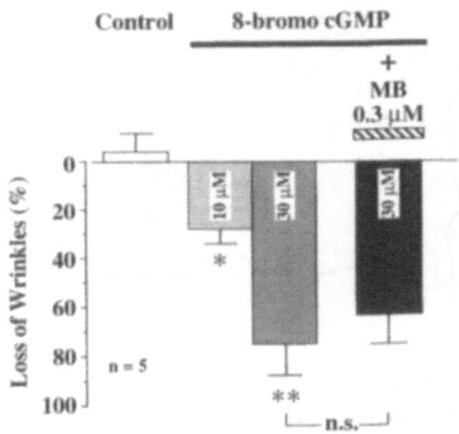

FIG. 19. Simplified drawing illustrates the fact that the cGMP analog 8-bromo-cGMP relaxes retinal pericytes

FIG. 20. Simplified drawing illustrates the NO–guanylate cyclase pathway in retinal pericytes

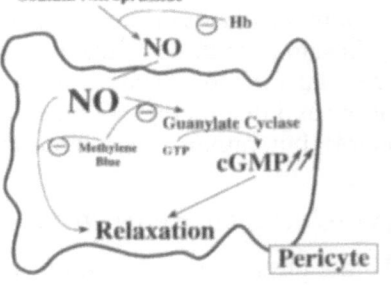

Pericytes were relaxed with either (3 and 10 μM) sodium nitroprusside or (10 μM) forskolin, drugs that stimulate the guanylate cyclase and adenylate cyclase pathways, respectively (Ferrari-Dileo et al. 1992; Haefliger et al. 1994b). Experiments were conducted under hyperoxic and hypoxic conditions. Relaxation induced by sodium nitroprusside (3 and 10 μM) was significantly more pronounced in the hypoxic than in the hyperoxic environment (Fig. 21). In contrast,

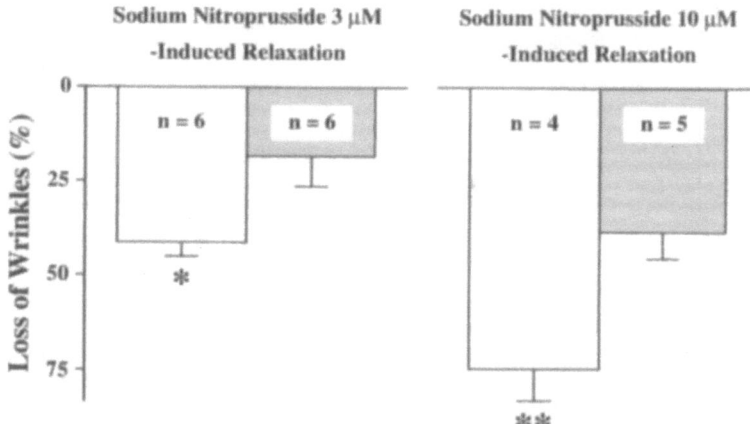

FIG. 21. Effect of 100% oxygen (hyperoxia) (*shaded bars*) and 100% nitrogen (hypoxia) (*open bars*) on the relaxation evoked by 3 and 10 μM sodium nitroprusside. Relaxation was more pronounced during hypoxia than during hyperoxia. *$P < 0.05$; **$P < 0.01$. (Data from Haefliger and Anderson 1997a)

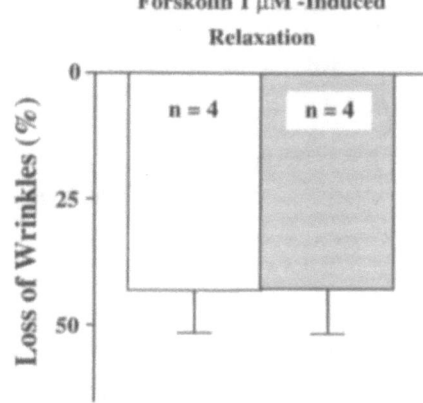

FIG. 22. Effect of 100% oxygen (hyperoxia) (*Shaded bar*) and 100% nitrogen (hypoxia) (*open bar*) on the relaxation evoked by 1 μM forskolin, an activator of adenylate cyclase. Under hypoxic and in hyperoxic conditions, the relaxation induced by forskolin was similar (Haefliger and Anderson 1997a)

no significant difference could be detected between the relaxations induced by forskolin 1 μM under hypoxic and hyperoxic conditions (Fig. 22).

SIN-1- and ANP-Induced Relaxation

To further address this issue the relaxing effect of two other drugs were studied: SIN-1 and ANP. Like sodium nitroprusside, these two substances could also relax pericytes in a concentration-dependent manner with an ER_{50} of 0.1 μM (Figs. 23, 24).

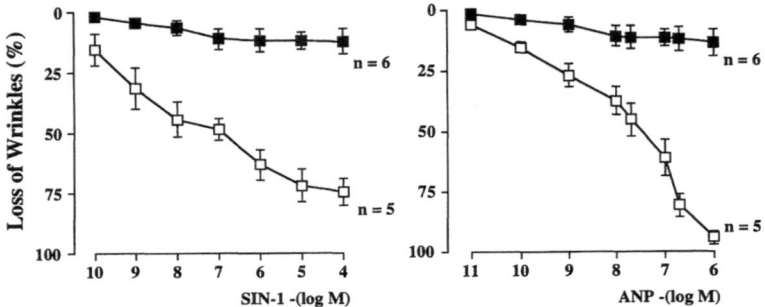

FIG. 23. Concentration response curve showing the relaxing effect of 3-morpholino syd-nonimine (*SIN-1*) and atrial natriuretic peptide (ANP) on the tone of retinal pericytes. Time-control experiments (*filled squares*). *Open squares*, controls. (From Haefliger and Anderson 1997b, with permission)

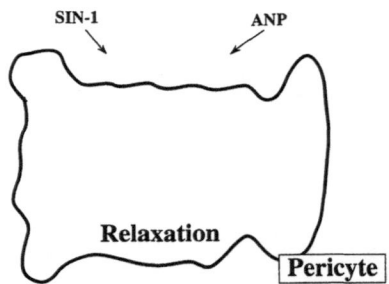

FIG. 24. Simplified drawing illustrates the fact that SIN-1 and ANP relax retinal pericytes

SIN-1 and ANP Stimulation of Guanylate Cyclase

The relaxations induced by both SIN-1 and ANP were mediated by production of cGMP through the stimulation of the enzyme guanylate cyclase. Both relaxations were completely inhibited by the inhibitor of guanylate cyclase, methylene blue (Figs. 25, 26).

NO Mediation of SIN-1-Induced Relaxation

To determine if activation of the guanylate cyclase pathway by SIN-1 and ANP was mediated by the release of NO, cells were exposed to SIN-1 and ANP in the absence and presence of hemoglobin, a scavenger of NO. The SIN-1-induced relaxation was completely abolished by hemoglobin, but the relaxation evoked by ANP was not, which is best explained by the fact that SIN-1 releases

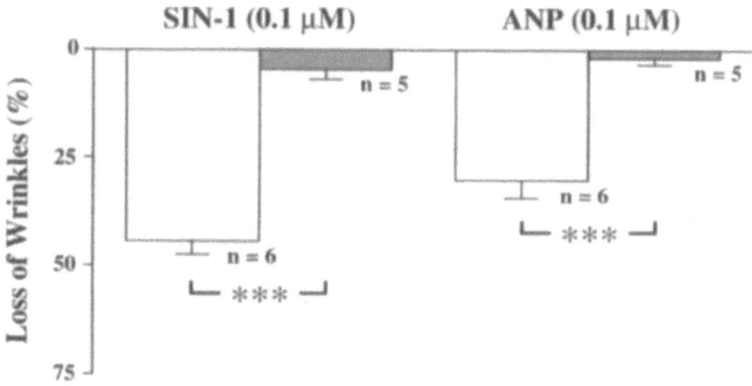

FIG. 25. Inhibitory effect of methylene blue (MB), an inhibitor of guanylate cyclase, on the relaxation evoked by SIN-1 and ANP, respectively. *Open bars*, without MB; *shaded bars*, with MB. ***$P < 0.001$. (From Haefliger and Anderson 1997b, with permission)

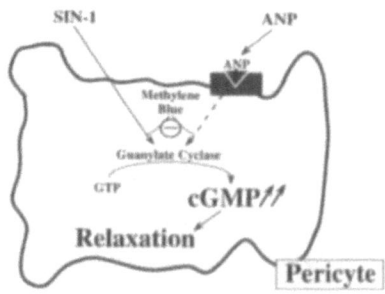

FIG. 26. Simplified drawing illustrates the fact that methylene blue inhibits relaxation evoked by SIN-1 and ANP

NO (Figs. 27, 28), whereas ANP relaxes pericytes by activating a guanylate cyclase bound to a membrane receptor (Leitman et al. 1988).

Indeed, ANP is known to stimulate guanylate cyclase after binding to a membrane receptor, not through the release of NO. In bovine retinal microvascular pericytes, in contrast to arterial smooth muscle cells (Rapoport et al. 1985), the effect of ANP was inhibited by methylene blue. Such inhibition has been shown in various cells (Kurz et al. 1986; de Venes et al. 1988; Henrich et al. 1988; Tamaoki et al. 1991), even in vascular smooth muscle cells from the lymphatic system (Ohhashi 1990). Inhibition of ANP-induced guanylate cyclase stimulation by methylene blue has been attributed to the existence of a specific methylene blue-sensitive ANP membrane receptor subtype (Fethiere et al. 1989; Shigematsu et al. 1993). The present observations in pericytes show that within the cardiovascular system, in addition to the known heterogeneity in the vascular responses to ANP (it relaxes arteries but not veins) (Ignarro et al. 1986), a heterogeneity

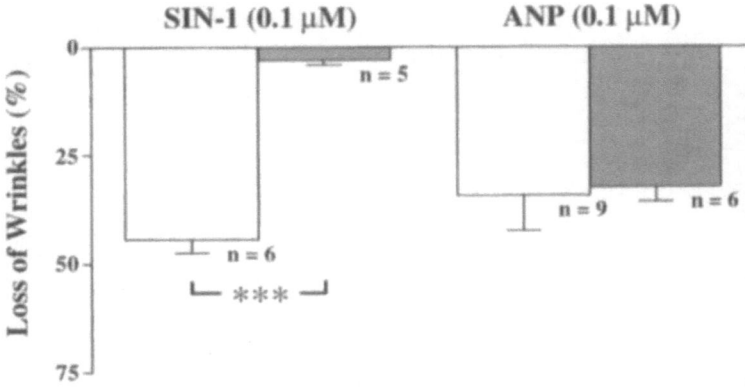

FIG. 27. Inhibitory effect of hemoglobin, a scavenger of NO, on the relaxation evoked by SIN-1 but not that induced by ANP. *Open bars*, without hemoglobin; *shaded bars*, with hemoglobin 10 μM. ***$P < 0.001$. (From Haefliger and Anderson 1997b, with permission)

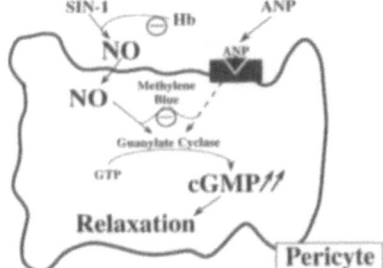

FIG. 28. Simplified drawing illustrates the fact that hemoglobin inhibits relaxations evoked by SIN-1

could also exist in the distribution of different ANP-receptor subtypes (Fethiere et al. 1989).

In other words, these two last experiments demonstrate that both SIN-1 and ANP relax pericytes through stimulation of the guanylate cyclase pathway. SIN-1, however, stimulates a soluble guanylate cyclase through the release of NO, whereas ANP most likely accomplishes it through activation of a membrane receptor.

Effect of Oxygen on Relaxation Evoked by SIN-1 and ANP

The relaxation evoked by SIN-1 was increased under hypoxic conditions compared to normoxic conditions, whereas under hyperoxic conditions the relaxation was significantly decreased. This was not the case with ANP, although this

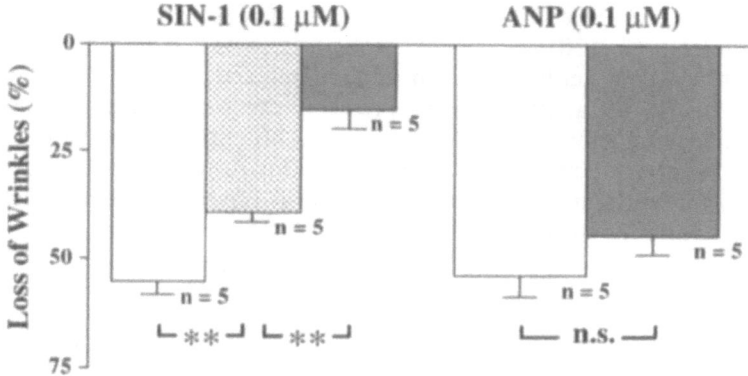

FIG. 29. Effect of different concentrations of oxygen on the relaxations evoked by SIN-1 and ANP. Only relaxations induced by SIN-1 (mediated by extracellular NO) were modulated by oxygen, but not those evoked by ANP. *Open bars*, hypoxia (100% N_2); *dotted bars*, normoxia (ambient air); *shaded bars*, hyperoxia (100% O_2). **$P < 0.01$; n.s., not significant. (From Haefliger and Anderson 1997b, with permission)

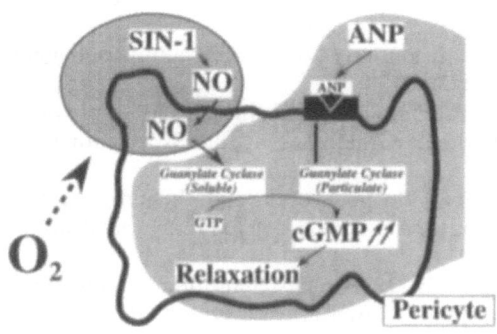

FIG. 30. Simplified drawing illustrates the fact that oxygen modulates relaxation mediated by the activation of the guanylate cyclase pathway most likely through an interaction with NO

drug stimulates the guanylate cyclase pathway (Figs. 29, 30). The results of this experiment demonstrate that pericyte relaxation can be modulated by various concentrations of oxygen when the relaxation is evoked by NO.

Conclusions

The results demonstrate that SIN-1 and ANP relax pericytes by activating the enzyme guanylate cyclase, and that SIN-1 relaxes pericytes by releasing extracellular NO. Oxygen modulates the relaxation to SIN-1 most likely by an interaction between oxygen and NO and not by interfering with the relaxing properties of the cells (Fig. 30).

Pericytes are contractile cells that can be relaxed by NO (Haefliger and Anderson 1996b, 1997a, 1997b). They probably can modify the diameter of a retinal capillary (Wallow et al. 1993) and thus potentially flow within such a vessel (Tilton et al. 1979; Voyata et al. 1984). Our results suggest that oxygen could participate, by an interaction with NO, in physiological modulation of blood flow within the capillary network.

References

Anderson DR (1970) Vascular supply to the optic nerve of primates. Am J Ophthalmol 60:341–351

Anderson DR (1989) The damage caused by pressure. Am J Ophthalmol 108:485–495

Anderson DR (1996) Glaucoma, capillaries and pericytes. 1. Blood flow regulation. Ophthalmologica 210:257–262

Anderson DR, Braverman S (1976) Reevaluation of the optic disc vasculature. Am J Ophthalmol 82:165–174

Anderson DR, Davis EB (1996) Glaucoma, capillaries, and pericytes. 2. Identification and characterization of retinal pericytes in culture. Ophthalmologica 210:263–268

Anderson DR, Quigley HA (1992) The optic nerve. In: Hart WM Jr (ed) Alder's physiology of the eye. 9th ed. Mosby-Year Book, St. Louis, pp 616–640

Aotaki-Keen AE, Harvey AK, de Juan E, Hjelmand LM (1991) Primary culture of human retinal glia. Invest Ophthalmol Vis Sci 32:1733–1738

Béchetoille A, Bresson-Dumont H (1994) Diurnal and nocturnal blood pressure drops in patients with local ischemic glaucoma. Graefes Arch Clin Exp Ophthalmol 232:675–679

Benedito S, Prieto D, Nielsen PJ, Nyborg NCB (1991a) Role of the endothelium in acetylcholine-induced relaxation and spontaneous tone of bovine isolated retinal small arteries. Exp Eye Res 52:575–579

Benedito S, Prieto D, Nielsen PJ, Nyborg NCB (1991b) Histamine induces endothelium-dependent relaxation of bovine retinal arteries. Invest Ophthalmol Vis Sci 32:32–38

Bill A, Sperber GO (1990) Control or retinal and choroidal blood flow. Eye 4:319–325

Chakravarthy U, Gardiner TA, Anderson P, Archer DB, Trimble ER (1992) The effect of endothelin-1 on the retinal microvascular pericyte. Microvasc Res 43:241–254

Chan LS, Li W, Khatami M, Rockey JH (1986) Actin in cultured bovine retinal capillary pericytes: morphological and functional correlation. Exp Eye Res 43:41–54

D'Amore PA (1990) Culture and study of pericytes. In: Piepr HM (ed) Cell culture techniques in cardiovascular research. Springer, Heidelberg, pp 299–314

Das A, Frank RN, Weber ML, Kennedy A, Reidy CA, Mancini MA (1988) ATP causes retinal pericytes to contract in vitro. Exp Eye Res 46:349–362

Demailly P, Cambien F, Plouin F, Baron P, Chevallier B (1984) Do patients with low-tension glaucoma have particular cardiovascular characteristics? Ophthalmologica 188:65–75

De Venes J, Bol JG, Hudson L, Schipper J, Steinbusch HW (1988) Atrial natriuretic factor-responding and cyclic guanosine monophosphate and (cGMP)-producing cells in the rat hippocampus: a combined micropharmacological and immunocytochemical approach. Brain Res 446:387–395

Dodge AB, Hechtman HB, Shepro D (1991) Microvascular endothelial-derived autacoids regulate pericyte contractility. Cell Motil Cytoskeleton 18:180–188

Donati G, Pournaras CJ, Munoz JL, Poitry S, Poitry-Yamate CL, Tsacopoulos M (1995) Nitric oxide controls arteriolar tone in the retina of the miniature pig. Invest Ophthalmol Vis Sci 36:2228–2237

Fethiere J, Meloche S, Nguyen TT, Ong H, De Lean A (1989) Distinct properties of atrial natriuretic factor receptor subpopulations in epithelial and fibroblast cell line. Mol Pharmacol 35:584–592

Ferrari-Dileo G, Davis EB, Anderson DR (1992) Effects of cholinergic and adrenergic agonists on adenylate cyclase activity of retinal microvascular pericytes in culture. Invest Ophthalmol Vis Sci 33:42–47

Flammer J (1993) Therapeutical aspects of normal-tension glaucoma. Curr Opin Ophthalmol 4:58–64

Flammer J (1996) To what extent are vascular factors involved in the pathogenesis of glaucoma? In: Kaiser HJ, Flammer J, Hendrickson P (eds) Ocular blood flow. Karger, Basel, pp 12–39

Flammer J, Gasser P, Prünte Ch, Yao K (1992) The probable involvement of factors other than ocular pressure in the pathogenesis of glaucoma. In: Drance SM, Buskirk Van EM, Neufeld AH (eds) Pharmacology of glaucoma. Williams & Wilkins, Baltimore, pp 273–283

Frank RN, Dutta S, Mancini MA (1987) Pericyte coverage is greater in the retinal than in the cerebral capillaries of the rat. Invest Ophthalmol Vis Sci 28:1086–1091

Frank RN, Turczyn TJ, Das A (1990) Pericyte coverage of retinal and cerebral capillaries. Invest Ophthalmol Vis Sci 31:999–1007

Gasser P, Flammer J (1990) Short- and long-term effect of nifedipine on the visual field of patients with presumed vasospasm. J Int Med Res 18:334–339

Gasser P, Flammer J (1991) Blood-cell velocity in the nailfold capillaries of patients with normal-tension or high-tension glaucoma and of healthy controls. Am J Ophthalmol 111:585–588

Geijer A, Bill A (1979) Effects of raised intraocular pressure on retinal, prelaminar, laminar, and retrolaminar optic nerve blood flow in monkeys. Invest Ophthalmol Vis Sci 18:1030–1042

Graham SL, Drance SM, Wijsman K, Douglas GR, Mikelberg S (1995) Ambulatory blood-pressure monitoring in glaucoma. Ophthalmology 102:61–69

Grunwald JE, Riva CE, Stone RA, Keates EU, Petrig BL (1984) Retinal autoregulation in open-angle glaucoma. Ophthalmology 91:1690–1694

Guthauser U, Flammer J, Mahler F (1988) The relationship between digital and ocular vasospasm. Graefes Arch Clin Exp Ophthalmol 226:224–226

Guyton AC (1991) Overview of the circulation, and medical physics of pressure, flow, and resistance. In: Guyton AC (ed) Textbook of medical physiology. 8th ed. Saunders, Philadelphia, pp 150–157

Haefliger IO (1995) Regulation des Blutflusses in der Papille Search Glaucoma 3:80–85

Haefliger IO, Anderson DR (1996a) Blood flow regulation in the optic nerve head. In: Ritch R, Shields MB, Krupin T (eds) The glaucomas. 2nd ed. Mosby-Year Book, St. Louis, pp 189–197

Haefliger IO, Anderson DR (1996b) Pericytes and capillary blood flow modulation. In: Kaiser HJ, Flammer J, Hendrickson Ph (eds) Ocular blood flow. Karger, Basel, pp 74–78

Haefliger IO, Anderson DR (1997a) Effect of oxygen on relaxation of retinal pericytes by sodium nitroprusside. Graefes Arch Clin Exp Ophthalmol 235:388–392

Haefliger IO, Anderson DR (1997b) Oxygen modulation of guanylate cyclase-mediated retinal pericyte relaxations to SIN-1 and ANP. Invest Ophthalmol Vis Sci 38:1563–1568

Haefliger IO, Flammer J (1997a) The logic of the prevention of glaucomatous damage progression. Curr Opin Ophthalmol 8:35–36

Haefliger IO, Flammer J (1997b) Le syndrome vasospastique un facteur de risque associé au glaucome. In: Béchetoille A (ed) Glaucomes. 2nd ed. Jappernard, Nantes (273–275)

Haefliger IO, Flammer J, Lüscher TF (1992) Nitric oxide and endothelin-1 are important regulators of human ophthalmic artery. Invest Ophthalmol Vis Sci 33:2340–2343

Haefliger IO, Flammer J, Lüscher TF (1993a) Endothelium-derived factors as local modulators of the vascular tone: implications in the ophthalmic and cerebral circulation. In: Lehmenkühler A, Grotemeyer K-H, Tegtmeier D (eds) Migraine: basic mechanisms and treatment. Urban & Schwarzenberg, Munich, pp 185–202

Haefliger IO, Flammer J, Lüscher TF (1993b) Heterogeneity of endothelium-dependent regulation in ophthalmic and ciliary arteries. Invest Ophthalmol Vis Sci 34:1722–1730

Haefliger IO, Meyer P, Flammer J, Lüscher TF (1994a) The vascular endothelium as a regulator of the ocular circulation: a new concept in ophthalmology. Surv Ophthalmol 39:123–132

Haefliger IO, Zschauer A, Anderson DR (1994b) Relaxation of retinal pericytes contractile tone through the nitric oxide-cyclic guanosine monophosphate pathway. Invest Ophthalmol Vis Sci 35:991–997

Harris AK, Wild P, Stopak D (1980) Silicone rubber substrata: a new wrinkle in the study of cell locomotion. Science 208:177–179

Hayreh SS, Zimmerman BM, Podhajsky P, Alward WLM (1994) Nocturnal arterial hypotension and its role in optic nerve head and ocular ischemic disorders. Am J Ophthalmol 117:603–624

Helbig H, Kornacker S, Berweck S, Stahl F, Lepple-Wienhues A, Wiederholt M (1992) Membrane potentials in retinal capillary pericytes: excitability and effect of vasoactive substances. Invest Ophthalmol Vis Sci 33:2105–2112

Henrich WL, McAllister EA, Smith PB, Campbell WB (1988) Guanosine 3′,5′-cyclic monophosphate as a mediator of inhibition of renin release. Am J Physiol 255:F474–F478

Hoste AM, Andries LJ (1991) Contractile responses of isolated bovine retinal microarteries to acetylcholine. Invest Ophthalmol Vis Sci 32:1996–200

Ignarro LJ, Wood KS, Harbison RG, Kadowitz PJ (1986) Atriopeptin II relaxes and elevates cGMP in bovine pulmonary artery but not vein. J Appl Physiol 60:1128–1133

Joyce NC, DeCamilli P, Boyles J (1984) Pericytes, like vascular smooth muscle, contain high levels of cyclic GMP-dependent protein kinase. Microvasc Res 28:206–219

Joyce NC, Haire MF, Palade GE (1985a) Contractile proteins in pericytes. I. Immunoperoxidase localization of tropomyosin. J Cell Biol 100:1379–1386

Joyce NC, Haire MF, Palade GE (1985b) Contractile proteins in pericytes. II. Immunocytochemical evidence for the presence of two isomyosins in graded concentrations. J Cell Biol 100:1387–1395

Kaiser HJ, Flammer J (1991) Systemic hypotension: a risk factor for glaucomatous damage. Ophthalmologica 203:105–108

Kaiser HJ, Flammer J, Graf T, Stümpfig D (1993) Systemic blood pressure in glaucoma patients. Graefes Arch Clin Exp Ophthalmol 231:677–680

Kanellopoulos AJ, Erickson KA, Netland PA (1996) Systemic calcium channel blockers and glaucoma. J Glaucoma 5:357–362

Kelley C, D'Amore P, Hechtman HB, Shepro D (1987) Microvascular pericyte contractility in vitro: comparison with other cells of the vascular wall. J Cell Biol 104:483–490

Kelley C, D'Amore P, Hechtman HB, Shepro D (1988) Vasoactive hormones and cAMP affect pericyte contraction and stress fibers in vitro. J Muscle Res Cell Motil 9:184–194

Kitazawa J, Shirai H, Go FJ (1989) The effect of calcium antagonist on visual field in low-tension glaucoma. Graefes Arch Clin Exp Ophthalmol 227:408–412

Kurz A, Della Bruna R, Pfeilschifter J, Taugner R, Bauer C (1986) Atrial natriuretic peptide inhibits renin release from juxtaglomerular cells by a cGMP-mediated process. Proc Natl Acad Sci USA 83:4769–4773

Kuwabara T, Cogan DG (1963) Retinal vascular patterns. VI. Mural cells of the retinal capillaries. Arch Ophthalmol 69:492–502

Lee T-S, Hu K-Q, Chao T, King GL (1989) Characterization of endothelin receptors and effects of endothelin on diacylglycerol and protein kinase C in retinal capillary pericytes. Diabetes 38:1643–1646

Leitman DC, Andresen JW, Catalano RM, Waldman SA, Tuan JJ, Murad F (1988) Atrial natriuretic peptide binding, cross-linking, and stimulation of cyclic GMP accumulation and particulate guanylate cyclase activity in cultured cells. J Biol Chem 263:4720–4728

Lipowsky HH, Kolvalcheck S, Zweifach BW (1978) The distribution of blood rheological parameters in the microvasculature of cat mesentery. Circ Res 43:738–749

McLaren MJ, Inana G, Li CY (1993) Double fluorescent vital assay of phagocytosis by cultured retinal pigment epithelial cells. Invest Ophthalmol Vis Sci 34:317–326

Nayak RC, Berman AB, George KL, Eisenbarth GS, King GL (1988) A monoclonal antibody (3G5)-defined ganglioside antigen is expressed on the cell surface of microvascular pericytes. J Exp Med 167:1003–1015

Ohhashi T, Watanabe N, Kawai Y (1990) Effect of atrial natriuretic peptide on isolated bovine mesenteric lymph vessels. Am J Physiol 259:H42–47

Petrig B, Werner EB, Riva CE, Grunwald J (1985) Response of macular capillary blood flow to changes in intraocular pressure as measured by blue field stimulation technique. Doc Ophthalmol Proc Ser 42:447–451 (Sixth International Visual Field Symposium)

Pillunat LE (1998) Vasoactive stimuli and visual field stimulation. In: Haefliger IO, Flammer J (eds) NO and endothelin in the pathogenesis of glaucoma. Lippincott-Raven, New York, pp 89–101

Pillunat LE, Stodtmeister R, Wilmanns I, Christ T (1985) Autoregulation of ocular blood flow during changes in intraocular pressure: preliminary results. Graefes Clin Exp Ophthalmol 223:219–223

Pillunat LE, Stodtmeister R, Wilmanns I (1987) Pressure compliance of the optic nerve head in low tension glaucoma. Br J Ophthalmol 71:181–187

Pillunat LE, Lang GK, Harris A (1994) The visual response to increased ocular blood flow in normal pressure glaucoma. Surv Ophthalmol 38(Suppl):S139–S148

Pillunat LE, Anderson DR, Knighton RW, Joos KM, Feuer WJ (1997) Autoregulation in human optic nerve head circulation in response to increased intraocular pressure. Exp Eye Res 64:737–744

Pournaras CJ (1996) Autoregulation of ocular blood flow. In: Kaiser HJ, Flammer J, Hendrickson Ph (eds) Ocular blood flow. Karger, Basel, pp 40–50

Rapoport RM, Waldman SA, Schwarta K, Winquist RJ, Murad F (1985) Effect of atrial natriuretic factor, sodium nitroprusside, and acetylcholine on cGMP levels and relaxation in rat aorta. Eur J Pharmacol 115:219–229

Riva CE, Grunwald JE, Sinclair SH (1983) Laser Doppler velocimetry of the effect of pure oxygen breathing on retinal blood flow. Invest Ophthalmol Vis Sci 24:47–51

Riva CE, Grunwald JE, Petrig BL (1986) Autoregulation of human retinal blood flow: an investigation with laser Doppler velocimetry. Invest Ophthalmol Vis Sci 27:1706–1712

Robert Y, Steiner D, Hendrickson P (1989) Papillary circulation dynamics in glaucoma. Graefes Arch Clin Exp Ophthalmol 227:436–439

Robinson R, Riva CE, Grunwald JE, Petrig BL, Sinclair SH (1986) Retinal blood flow autoregulation to an acute increase in blood pressure. Invest Ophthalmol Vis Sci 27:722–726

Shields MB (1992) Textbook of glaucoma. 3rd ed. Williams & Wilkins, Baltimore, pp 431–629

Shigematsu Y, Vaughn J, Touchard CL, Frohlich ED, Alam J, Cole FE (1993) Different ATP effects on natriuretic peptide receptor subtypes in LLC-PK1 and NIH-3T3 cells. Life Sci 53:865–874

Sinclair SH, Grunwald JE, Riva CE, Braunstein SN, Nichols CW, Schwarz SS (1982) Retinal vascular autoregulation in diabetes mellitus. Ophthalmology 89: 748–750

Sossi N, Anderson DR (1983) Effect of elevated intraocular pressure on blood flow: occurrence in cat optic nerve head studied with iodoantipyrine I-125. Arch Ophthalmol 101:98–101

Spaeth GL (1996) Proper outcome measurements regarding glaucoma: the inadequacy of using intraocular pressure alone. Eur J Ophthalmol 6:101–105

Tamaoki J, Kobayashi K, Sakai N, Kanemura T, Horii S, Isono K, Takeuchi S, Chiyotani A, Yamawaki I, Takizawa T (1991) Atrial natriuretic factor inhibits ciliary motility in cultured rabbit tracheal epithelium. Am J Physiol 260:C201–C205

Tilton RG, Kilo C, Williamson JR, Murch DW (1979) Differences in pericyte contractile function in rat cardiac and skeletal muscle microvasculatures. Microvasc Res 18: 336–352

Toussaint D, Kuwabara T, Cogan DG (1961) Retinal vascular patterns. II. Human retinal vessels studied in three dimensions. Arch Ophthalmol 65:575–581

Ulrich WD, Ulrich C, Bohne BD (1986) Deficient autoregulation and lengthening of the diffusion distance in the anterior optic nerve circulation in glaucoma: an electroencephalo-dynamographic investigation. Ophthalmol Res 18:253–259

Voyata J, Via D, Butterfield C, Zetter B (1984) Identification and isolation of endothelial cells based on their increased uptake of acetylated-low density lipoproteins. J Cell Biol 99:2034–2040

Wallow IH, Bindley CD, Reboussin DM, Gange SJ, Fisher MR (1993) Systemic hypertension produces pericyte changes in retinal capillaries. Invest Ophthalmol Vis Sci 34:420–430

Weinstein JM, Funsch D, Page RB, Brennan RW (1982) Optic nerve blood flow and its regulation. Invest Ophthalmol Vis Sci 23:640–645

Weinstein JM, Duckrow RB, Beard D, Brennan RW (1983) Regional optic nerve blood flow and its autoregulation. Invest Ophthalmol Vis Sci 24:1559–1565

Wink DA, Beckman JS, Ford PC (1996) Kinetics of nitric oxide reaction in liquid and gas phase. In: Freelish M, Stamler S (eds) Methods in nitric oxide research. Wiley, Chichester, pp 29–37

Zweifach BW, Lipowsky HH (1984) Pressure-flow relations in blood and lymph micro-circulation. In: Handbook of physiology. Sect 2: The cardiovascular system, Vol IV: Microcirculation. American Physiological Society, Bethesda, pp: 251–307

Regulation of Retinal Arterial and Arteriolar Tone by Nitric Oxide Derived from Endothelium and Perivascular Nerve

Noboru Toda[1,2], Megumi Toda[1,3], and Tomio Okamura[1]

Introduction

Nitric oxide (NO) is now recognized to be a key substance for regulating functions of not only the cardiovascular, nervous, and immune systems but also other organs and tissues of almost the entire body. The role of NO derived from the endothelium and perivascular nerve has not been well understood in the ocular circulation, possibly because detailed analyses in in vitro studies are difficult in such a small and fragile ocular artery. This chapter describes the role of endothelium-derived NO in isolated retinal central arteries and retinal arterioles in vivo. It also describes the role of NO derived from perivascular nerve in isolated retinal central, ophthalmic, and cerebral arteries and the retinal arterioles in vivo.

Endothelium-Derived Relaxing Factor and NO

Discovery of NO responsible for an important, endogenous mediator is based on the first, outstanding report on endothelium-derived relaxing factor (EDRF) by Furchgott and Zawadzki (1980). On the basis of findings that the vasodilator actions of acetylcholine are abolished or are reversed to contractions after removal of the endothelium in the isolated rabbit aorta, they hypothesized that acetylcholine liberates labile vasodilator substance(s) (EDRF) from the endothelium. Involvement of cyclooxygenase products was excluded. Acetylcholine is not the only substance that enables the release of EDRF; many other compounds, such as bradykinin, substance P, histamine, adenosine diphosphate

[1] Department of Pharmacology, Shiga University of Medical Science, Seta, Ohtsu 520-2192, Japan
[2] Toyama Institute for Cardiovascular Pharmacology Research, Toyama Bldg., Azuchi-machi, Chuo-ku, Osaka 541-0052, Japan
[3] Present address: Department of Ophthalmology, Osaka Medical College, Takatsuki, Osaka 569-8686, Japan

TABLE 1. Comparison of EDRF and nitric oxide

Parameter	EDRF	NO
Biological half-life	Very short (4–10 s in artificial solution)	Very short (4–10 s in artificial solution)
Stability		
Acid	Stable	Stable
Alkali	Unstable	Unstable
Lyophilization	Stable	Stable
Bound to column		
Anion exchange	Yes	No
Hydrophobic (C18)	No	No
Release from endothelial cells	Yes	Yes
Relaxation		
Nonvascular smooth muscle	Yes (weak) or no	Yes
Vascular smooth muscle	Yes	Yes
Modification by		
Antioxidants	Inhibition	Inhibition
SOD	Potentiation	Potentiation
Oxyhemoglobin	Inhibition	Inhibition
Soluble guanylate cyclase	Activation	Activation
Second messenger	cGMP	cGMP
Platelet aggregation/ adhesion	Inhibition	Inhibition

SOD, superoxide dismutase; EDRF, endothelium-derived relaxing factor; NO, nitric oxide; cGMP, cyclic guanosine monophosphate.

(ADP), and thrombin, have been known to elicit relaxation via mediation of EDRF (Angus and Cocks 1989). The characteristics of EDRF and NO are summarized in Table 1.

It was independently reported from three research groups (Ignarro et al. 1987; Palmer et al. 1987; Furchgott 1988) that the biological and chemical natures of EDRF are similar to those of NO, as indicated in Table 1. Palmer et al. (1988a) detected labeled NO in culture media that contained [^{15}N]-L-arginine, in which incubated endothelial cells were stimulated by bradykinin. L-Arginine analogues, such as N^G-monomethyl-L-arginine (L-NMMA) and N^G-nitro-L-arginine (L-NA), depress the production of NO (Palmer et al. 1988b; Moore et al. 1990; Toda et al. 1990). The synthesis, release, and action of endothelium-derived NO are illustrated in Fig. 1. Vasodilator responses resistant to NO synthase inhibitors have also been reported (Enokibori et al. 1994). Non-NO EDRF is believed to produce relaxation by opening K^+ channels of smooth muscle cell membrane, resulting in hyperpolarization. Thus the substance(s) liberated from the endothelium is called endothelium-derived hyperpolarizing factor (Suzuki and Chen 1990).

Fig. 1. Transduction and smooth muscle action of endothelium-derived nitric oxide (NO). *R*, receptor; *L-NMMA*, *N*ᴳ-monomethyl-ʟ-arginine; *L-NA*, *N*ᴳ-nitro-ʟ-arginine; *EDRF*, endo-thelium-derived relaxing factor; *RNO*, NO analog such as *S*-nitrosothiol; *MB*, methylene blue; *sGC*, soluble guanylate cyclase; .*O*₂⁻, superoxide anion; *SOD*, superoxide dismu-tase; *OxyHb*, oxyhemo-globin; *NTG*, nitroglycerin; *GTP*, guano-sine triphos-phate; *cGMP*, 3′,5′-cyclic guanosine monophos-phate. *Bold arrows* indicate the inhibitory action

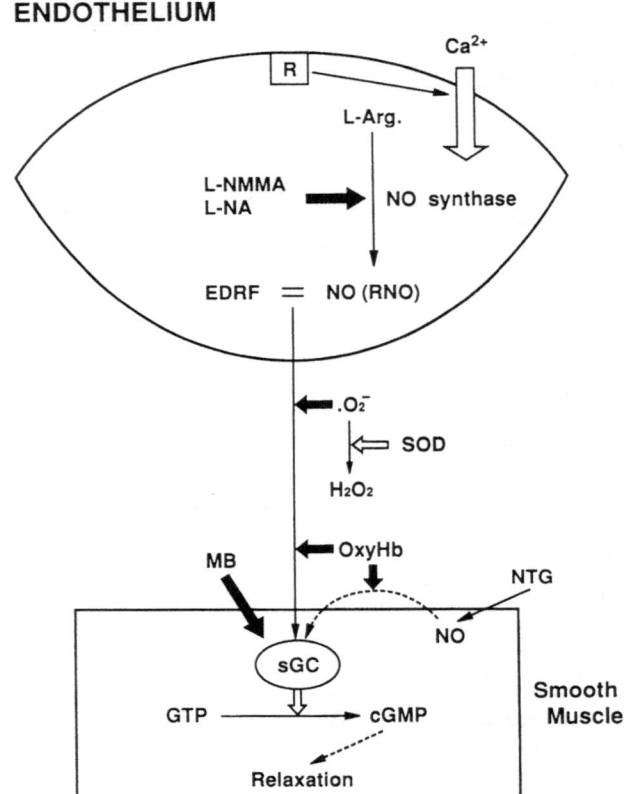

NO Derived from Endothelium

Endothelium-derived NO is synthesized from ʟ-arginine by catalysis of consti-tutive NO synthase, which is activated by Ca^{2+} intracellularly introduced due to stimulation of drug receptors on endothelial cell membrane or mechanical stim-ulation such as shear stress (Moncada et al. 1991; Dusting 1995). NO luminally and abluminally liberated from the endothelium plays important physiological roles, including vasodilatation and inhibition of platelet aggregation and adhe-sion, smooth muscle cell proliferation, and oxidation of low-density lipoprotein. Therefore vasoconstriction is minimized, vascular resistance is decreased, throm-bosis is protected, and atherosclerosis is prevented.

Abluminally released NO permeates the smooth muscle membrane to reach an intracellular target, soluble guanylate cyclase; and the activated enzyme increases the production of cyclic guanosine monophosphate (cGMP) from guanosine triphosphate (GTP) (Fig. 1). Extracellularly liberated NO is inactivated by superoxide anions (Gryglewski et al. 1986) and oxyhemoglobin (Martin et al. 1985). Superoxide dismutase potentiates and prolongs the action of NO. The guanylate cyclase activity is inhibited by methylene blue (Gruetter et al. 1981) or 1H-[1,2,4]oxadiazolo[4,3-a]quinoxaline-1-one (ODQ) (Brunner et al. 1996), which has more selective effects on the enzyme than methylene blue.

Responsiveness of Central Retinal and Other Ocular Arteries

Substance P in concentrations ranging from 10^{-10} to 10^{-8}M produces dose-dependent relaxation in isolated canine central retinal arterial strips partially contracted with prostaglandin (PG) $F_{2\alpha}$. Endothelium denudation abolishes peptide-induced relaxation, suggesting that the response is entirely endothelium-dependent (Kitamura et al. 1993). Prostanoids are not involved, because the response is not reduced by treatment with cyclooxygenase inhibitors. Similar findings were also observed with isolated canine external and internal ophthalmic arteries (Wang et al. 1993). The peptide-induced relaxation in endothelium-intact retinal arteries is dose-dependently suppressed by L-NA (10^{-7} and 10^{-6}M) (Fig. 2) but is not influenced by the D-enantiomer. L-Arginine, but not D-arginine, restores the response (Kitamura et al. 1993). However, relaxation induced by exogenous NO (acidified NaNO$_2$ solution) (Furchgott 1988) or nitroglycerin is not inhibited by the NO synthase inhibitor. The responses to substance P are abolished by treatment with methylene blue and oxyhemoglobin, as are those to NO and nitroglycerin, suggesting the involvement of cGMP also in the peptide-induced relaxation of retinal arteries.

It has been demonstrated that endothelium-derived NO participates in the response of isolated bovine retinal and porcine ophthalmic arteries to acetylcholine (Benedito et al. 1991; Yoo et al. 1991). However, acetylcholine-induced relaxation of canine retinal arteries is endothelium-independent and mediated by vasodilator prostanoids (Toda et al. 1995a,b). Bradykinin and histamine are reported to vasodilate isolated human ophthalmic arteries (Hoefliger et al. 1992). The response is depressed by treatment with NO synthase inhibitors—hence our speculation about the involvement of NO. Whether the response is dependent on the endothelium has not been determined.

Responsiveness of Arterioles in Ocular Fundus

Arterioles and venules in the ocular fundus of anesthetized dogs, have been recorded on video by fundus camera. Injections of "norepinephrine into the carotid artery vasoconstrict the arterioles, whereas substance P injections elicit

FIG. 2. Typical recordings of the responses to substance P (*SP*, 10^{-8}M), nitric oxide (*NO*, 10^{-7}M), and nitroglycerin (*NTG*, 10^{-8}M) of a dog retinal artery strip with intact endothelium. Responses of the strip were obtained before (control) and after treatment with N^{G}-nitro-L-arginine (*L-NA*) and 10^{-5}M L-NA plus L-arginine (*L-arg.*, 10^{-3}M). The strip was pretreated with 10^{-6}M indomethacin and partially contracted with prostaglandin $F_{2\alpha}$ ($PGF_{2\alpha}$). PA represents 10^{-4}M papaverine, which produces maximal relaxation. (From Kitamura et al. 1993, with permission)

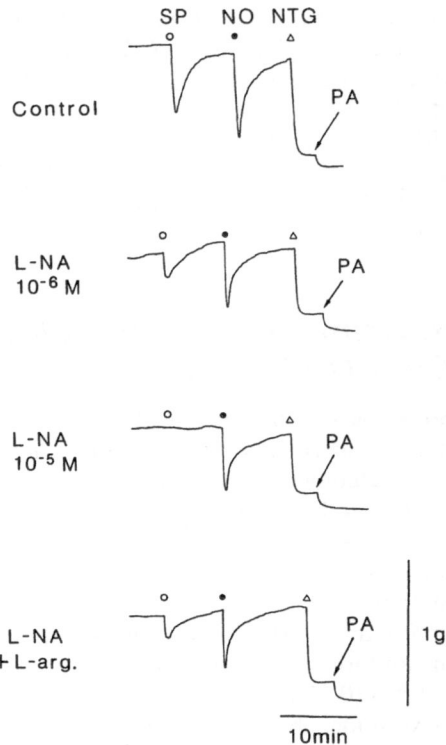

vasodilatation (Kitamura et al. 1993). The maximal response to the peptide (in a dose of 1.5 ng/kg) is attained 15 s after the injection; systemic blood pressure starts to lower 30 s later, and the maximal depressor response is seen about 40 s after that (Fig. 3). The blood pressure change does not seem to influence the observed retinal arteriolar dilatation induced by the peptide. Intravenous application of L-NA markedly inhibits or abolishes the vasodilator effect of substance P but does not alter the response to nitroglycerin in a dose sufficient to produce a similar degree of vasodilatation (Kitamura et al. 1993). L-NA-induced inhibition is antagonized by intravenous L-arginine. It is thus concluded that the retinal arteriolar dilatation due to substance P is mediated by NO. Although the endothelium dependence is not determined in the in vivo study, findings on isolated retinal arteries in response to the peptide suggest that NO is derived from the endothelium in retinal arterioles. This was the first demonstration of the response of arterioles to endogenous NO in vivo without surgical invasion, such as craniotomy and laparotomy.

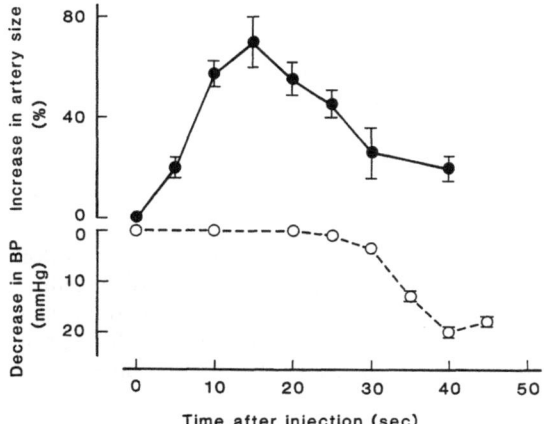

FIG. 3. Time course of changes by substance P (1.5 ng/kg) of retinal arteriolar size (*filled circles, top curve*) and mean blood pressure (*BP*) in anesthetized dogs (*open circles, bottom curve*). The BP before the drug injection averaged 77 ± 13 mmHg (*n* = 4). Vasodilatation induced by papaverine was taken as 100%. Vertical bars represent the standard error of the mean (SEM). (From Kitamura et al. 1993, with permission).

NO Derived from Perivascular Nerve

About 20 years ago we first found that nicotine or transmural electrical stimulation produces relaxation of isolated canine and monkey cerebral arterial strips and is not inhibited by treatment with atropine and β-adrenoceptor antagonists (Toda 1975, 1982). Histochemical studies have demonstrated the presence of perivascular nerve fibers containing polypeptides, such as vasoactive intestinal peptide (VIP), substance P, and calcitonin gene-related peptide (CGRP), in cerebral arteries from a variety of mammals (Owman 1990). These potent cerebral vasodilator peptides are believed to be candidates for neurotransmitters in the nonadrenergic, noncholinergic (NANC) nerve. However, our data on canine cerebral arteries desensitized to these peptides (e.g., VIP) (Fig. 4) by treatment with high concentrations of the peptides do not support this idea (Toda 1982; Okamura et al. 1989; Toda and Okamura 1991a,b; Toda et al. 1996). Neurogenic vasodilatation is not reduced in the arteries desensitized to the peptides. Treatment with capsaicin, which depletes sensory neurotransmitters (Saito et al. 1989), does not interfere with the response to vasodilator nerve stimulation, a point that also excludes the possible involvement of substance P and CGRP (Okamura and Toda 1994). Because methylene blue and oxyhemoglobin abolish the response to vasodilator nerve stimulation, we have speculated that the neurogenic response is mediated by cGMP (Toda 1988). The mechanism was clarified when we added L-N^G-monomethyl arginine, a NO synthase inhibitor, to the preparation (Toda and Okamura 1990a).

Responsiveness of Isolated Retinal and Other Ocular Arteries

In isolated canine central retinal arteries contracted with PGF$_{2\alpha}$, transmural electrical stimulation or nicotine produces relaxation. The response to electrical stim-

CANINE MIDDLE CEREBRAL ARTERY-Transmural stimulation

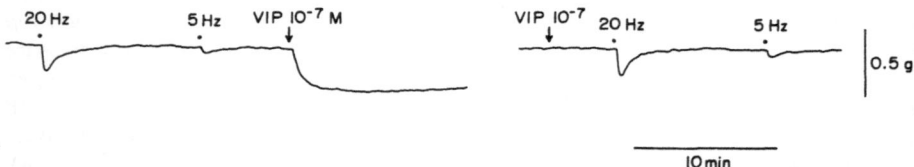

FIG. 4. Relaxant responses to transmural electrical stimulation (5 and 20 Hz) of a dog middle cerebral artery strip before and after desensitization to vasoactive intestinal polypeptide (*VIP*) by repeated applications of the peptide. The strip was partially contracted with $PGF_{2\alpha}$. The *horizontal line just left of the left tracing* represents the level prior to the application of $PGF_{2\alpha}$. In the *right tracing*, electrical stimulation was applied after abolishment of the response to VIP was confirmed.

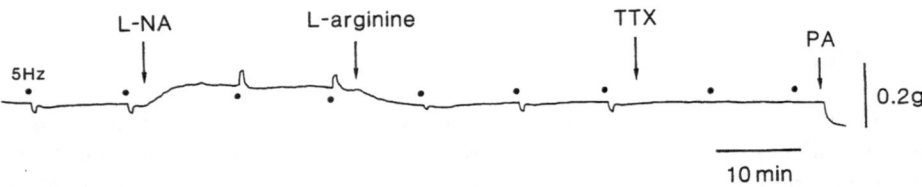

FIG. 5. Modification by N^G-nitro-L-arginine (*L-NA*, 10^{-6} M) and L-arginine (3×10^{-4} M) of the relaxation induced by transmural electrical stimulation at 5 Hz in a dog retinal arterial strip contracted with $PGF_{2\alpha}$. Tetrodotoxin (*TTX*, 3×10^{-7} M) abolished the response. *PA*, 10^{-4} M papaverine

ulation is abolished by tetrodotoxin, and the nicotine-induced relaxation is abolished by hexamethonium or other ganglionic blocking agents, suggesting involvement of vasodilator nerve activation. The responses are not influenced by atropine, timolol, or indomethacin or by endothelial denudation; they are abolished by L-NA, oxyhemoglobin, and methylene blue. The relaxation abolished by L-NA is restored by L-arginine, but not D-arginine (Toda et al. 1993, 1994). A typical recording is illustrated in Fig. 5; the neurogenic contraction seen under treatment with L-NA is abolished by prazosin, suggesting the response to be induced by adrenergic nerve stimulation. Thus the endothelium-independent relaxation associated with nerve stimulation is believed to be mediated by NO and cGMP. Relaxation induced by exogenous NO and sodium nitroprusside is not reduced by L-NA but is abolished by methylene blue and hemoglobin. In superfused cerebral arterial strips denuded of endothelium, the release of NO, measured as NO_x in the superfusate, and the content of cGMP in the tissue are evidently increased during nerve stimulation by electrical pulses or nicotine (Toda and Okamura 1990b, 1991a,b). The effect is abolished by treatment with NO synthase (NOS) inhibitors. Histochemical studies with NOS antibody or

NADPH-diaphorase method demonstrate the presence of abundant positively stained nerve fibers around the central retinal artery in the portion used for the study on tension recordings or the biochemical study. These findings strongly support the hypothesis that NO acts as a vasodilator neurotransmitter in canine retinal arteries. The nerve is called "nitroxidergic" (Toda and Okamura 1992a). The same conclusion has also been drawn from studies of internal and external ophthalmic arteries from dogs (Toda et al. 1995a,b). Human posterior ciliary arterial functions may also be regulated by nitroxidergic nerve (Nyborg and Nielsen 1994). Wiencke et al. (1994) suggested an involvement of NO and CGRP in neurogenic vasodilatation in bovine ciliary arteries.

Nitroxidergic innervation in central retinal arterial smooth muscle is summarized in Fig. 6. NO is synthesized from L-arginine by neuronal NOS, which is activated by Ca^{2+} introduced from extracellular fluids by nerve action potential or nicotine (Toda and Okamura 1992b; Toda et al. 1995a,b) in the presence of calmodulin (Okamura and Toda 1994). NO or its stable analog, such as S-nitrosothiol (RNO), liberated from the nerve activates soluble guanylate cyclase

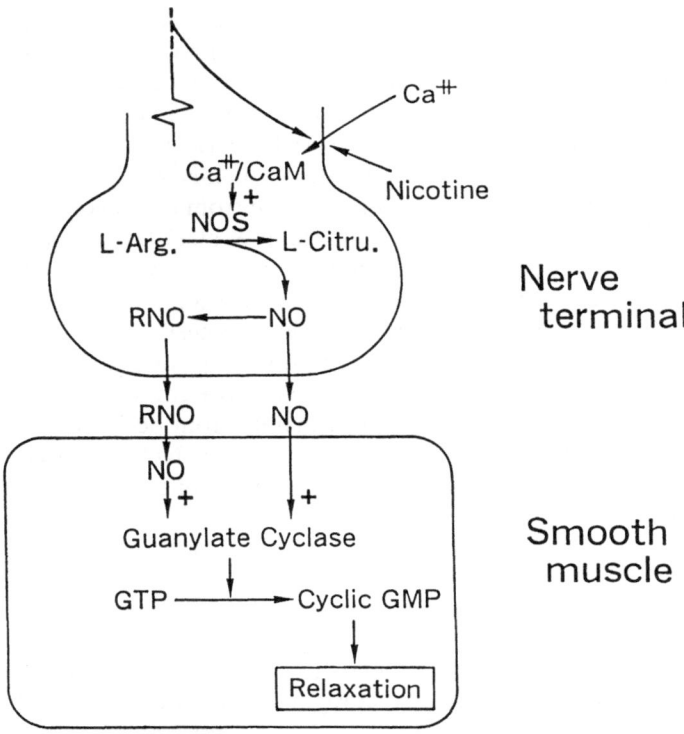

FIG. 6. Nitroxidergic innervation in the retinal artery. *NOS*, NO synthase; *CaM*, calmodulin; *L-Arg.*, L-arginine; *L-Citru.*, L-citrulline; *RNO*, stable analog of NO; +, stimulation

in smooth muscle cells and increases the production of cGMP, resulting in relaxation.

Responsiveness of Arterioles in Ocular Fundus

Nicotine injected into the carotid artery of anesthetized dogs vasodilates arterioles in the ocular fundus. The effect is diminished by treatment with L-NA or hexamethonium (Toda et al. 1994), as observed in isolated retinal arteries. L-Arginine antagonizes the inhibition by L-NA. It has been demonstrated histochemically that arterioles in the ocular fundus in dogs are innervated by nerve fibers containing NO synthase. These findings, together with the results from isolated central retinal arteries, led us to conclude that nicotine vasodilates the arterioles by mediation of NO liberated possibly from perivascular nerves.

Involvement of Endogenous NO in the Control of Ocular Blood Flow and Arteriolar Diameter In Vivo

The role of NO derived from perivascular nerve and endothelium in the control of ocular circulation in vivo can be evaluated by comparing the blood flow or arteriolar diameter before and after treatment with NOS inhibitors.

Blood flow in rabbit retinal, uveal (Seligsohn and Bill 1993), and choroidal vasculatures (Chiou et al. 1995) and cat choroidal (Mann et al. 1995) and newborn pig retinal (Gidday and Zhu 1995) vasculatures is decreased by intravenous injections or topical applications of L-NA, L-NA methylester, or L-monomethyl-arginine. Whether NO is released from the nerve or the endothelium has not been determined. Although Mann et al. (1995) suggested that NO involved is derived from the endothelium, there is no evidence supporting this hypothesis. Toda et al. (1994) demonstrated arteriolar constriction by intravenous injections of L-NA in the ocular fundus of anesthetized dogs that can be reversed by intravenous L-arginine. Based on findings of significant attenuation by hexamethonium of the L-NA-induced vasoconstriction, it is sugged that the NO involved in vasodilatation is derived mainly from perivascular nerve. This suggestion does not exclude the possibility that NO liberated spontaneously from the endothelium under resting conditions participates in retinal vasodilatation.

Conclusion

It is possible that NO synthesized from L-arginine in the perivascular nerve and endothelium plays an important role in regulating the ocular circulation, including blood flow in the retina and adjacent tissues. Impairment of the NO–cGMP system may participate in reduced ocular blood flow in association with severe vasoconstriction and thrombosis.

References

Angus JA, Cocks TM (1989) Endothelium-derived relaxing factor. Pharmacol Ther 41:303–351

Benedito S, Prieto D, Nielsen PJ, Nyborg NCB (1991) Role of the endothelium in acetyl-choline-induced relaxation and spontaneous tone of bovine isolated retinal small arteries. Eye Exp Res 52:575–579

Brunner F, Schmidt K, Nielsen EB, Mayer B (1996) Novel guanylyl cyclase inhibitor potently inhibits cyclic GMP accumulation in endothelial cells and relaxation of bovine pulmonary artery. J Pharmacol Exp Ther 277:48–53

Chiou GCY, Liu SXL, Li BHP, Chiang CH, Varma RS (1995) Ocular hypotensive effects of L-arginine and its derivatives and their actions on ocular blood flow. J Ocul Pharmacol Ther 11:1–10

Dusting GJ (1995) Nitric oxide in cardiovascular disorders. J Vasc Res 32:143–161

Enokibori M, Okamura T, Toda N (1994) Mechanism underlying substance P-induced relaxation in dog isolated superficial temporal arteries. Br J Pharmacol 111: 77–82

Furchgott RF (1988) Studies on relaxation of rabbit aorta by sodium nitrite: the basis for the proposal that the acid-activatable inhibitory factor from bovine retractor penis is inorganic nitrite and the endothelium-derived relaxing factor is nitric oxide. In: Vanhoutte PM (ed) Mechanism of vasodilatation. Vol IV. Lippincott-Raven, New York, pp 401–414

Furchgott RF, Zawadzki (1980) The obligatory role of endothelial cells in the relaxation of arterial smooth muscle by acetylcholine. Nature 288:373–376

Gidday JM, Zhu Y (1995) Nitric oxide does not mediate autoregulation of retinal blood flow in newborn pig. Am J Physiol 269:H1065–H1072

Gruetter CA, Kadowitz PJ, Ignarro LJ (1981) Methylene blue inhibits coronary arterial relaxation and guanylate cyclase activation by nitroglycerin, sodium nitrite, and amyl nitrite. Can J Physiol Pharmacol 59:150–156

Gryglewski RJ, Palmer RMJ, Moncada S (1986) Superoxide anion is involved in the breakdown of endothelium-derived relaxing factor. Nature 320:454–456

Hoefliger IO, Flammer J, Lüscher TF (1992) Nitric oxide and endothelin-1 are important regulators of human ophthalmic artery. Invest Ophthalmol Vis Sci 33:2340–2343

Ignarro LJ, Byrns RE, Buga GM, Wood KS (1987) Endothelium-derived relaxing factor from pulmonary artery and vein possesses pharmacologic and chemical properties identical to those of nitric oxide radical. Circ Res 60:82–92

Kitamura Y, Okamura T, Kani K, Toda N (1993) Nitric oxide-mediated retinal arteriolar and arterial dilatation induced by substance P. Invest Ophthalmol Vis Sci 34:2859–2865

Mann RM, Riva CE, Stone RA, Barnes GE, Cranstoun SD (1995) Nitric oxide and choroidal blood flow regulation. Invest Ophthalmol Vis Sci 36:925–930

Martin W, Villani GM, Jothianandan D, Furchgott RF (1985) Blockade of endothelium-dependent and glyceryl trinitrate-induced relaxation of rabbit aorta by certain ferrous hemoproteins. J Pharmacol Exp Ther 233:679–685

Moncada S, Palmer RMJ, Higgs A (1991) Nitric oxide: physiology, pathophysiology, and pharmacology. Pharmacol Rev 43:109–142

Moore PK, Al-Swayeh OA, Chong NWS, Evans RA, Gibson A (1990) L-N^G-Nitroarginine (L-NOARG), a novel, L-arginine-reversible inhibitor of endothelium-dependent vasodilatation in vitro. Br J Pharmacol 99:408–412

Nyborg NCB, Nielsen PJ (1994) Neurogenic nitric oxide accounts for the non-adrenergic non-cholinergic vasodilation in human posterior ciliary arteries. Invest Ophthalmol Vis Sci 35:1287 (abstract)

Okamura T, Toda N (1994) Mechanism underlying nicotine-induced relaxation in dog saphenous arteries. Eur J Pharmacol 263:85–91

Okamura T, Inoue S, Toda N (1989) Action of atrial natriuretic peptide (ANP) on dog cerebral arteries: evidence that neurogenic relaxations is not mediated by release of ANP. Br J Pharmacol 97:1258–1264

Owman C (1990) Peptidergic vasodilator nerves in the peripheral circulation and in the vascular beds in the heart and brain. Blood Vessels 27:73–93

Palmer RMJ, Ferrige AG, Moncada S (1987) Nitric oxide release accounts for the biological activity of endothelium-derived relaxing factor. Nature 327:524–526

Palmer RMJ, Ashton DS, Moncada S (1988a) Vascular endothelial cells synthesize nitric oxide from L-arginine. Nature 333:664–666

Palmer RMJ, Rees DD, Ashton DS, Moncada S (1988b) L-Arginine is the physiological precursor for the formation of nitric oxide in endothelium-dependent relaxation. Biochem Biophys Res Commun 153:1251–1256

Saito A, Masaki T, Uchiyama Y, Lee JFT, Goto K (1989) Calcitonin gene-related peptide and vasodilator nerves in large cerebral arteries of cats. J Pharmacol Exp Ther 248:455–462

Seligsohn EE, Bill A (1993) Effects of N^G-nitro-L-arginine methylester on the cardiovascular system of the anaesthetized rabbit and on the cardiovascular response to thyrotropin-releasing hormone. Br J Pharmacol 109:1219–1225

Suzuki H, Chen G (1990) Endothelium-derived hyperpolarizing factor (EDHF): an endogenous potassium-channel activator. News Physiol Sci 5:212–215

Toda N (1975) Nicotine-induced relaxation in isolated canine cerebral arteries. J Pharmacol Exp Ther 193:376–384

Toda N (1982) Relaxant responses to transmural stimulation and nicotine of dog and monkey cerebral arteries. Am J Physiol 243:H145-H153

Toda N (1988) Hemolysate inhibits cerebral artery relaxation. J Cereb Blood Flow Metab 8:46–53

Toda N, Okamura T (1990a) Modification by L-N^G-monomethyl arginine (L-NMMA) of the response to nerve stimulation in isolated dog mesenteric and cerebral arteries. Jpn J Pharmacol 52:170–173

Toda N, Okamura T (1990b) Possible role of nitric oxide in transmitting information from vasodilator nerve to cerebroarterial muscle. Biochem Biophys Res Commun 170:308–313

Toda N, Okamura T (1991a) Suppression by N^G-monomethyl-L-arginine of cerebroarterial responses to nonadrenergic, noncholinergic vasodilator nerve stimulation. J Cardiovasc Pharmacol 17(Suppl 3):S234–S237

Toda N, Okamura T (1991b) Role of nitric oxide in neurally induced cerebroarterial relaxation. J Pharmacol Exp Ther 258:1027–1032

Toda N, Okamura T (1992a) Regulation by nitroxidergic nerve of arterial tone. News Physiol Sci 7:148–152

Toda N, Okamura T (1992b) Different susceptibility of vasodilator nerve, endothelium and smooth muscle functions to Ca^{++} antagonists in cerebral arteries. J Pharmacol Exp Ther 261:234–239

Toda N, Minami Y, Okamura T (1990) Inhibitory effects of L-N^G- nitroarginine on the synthesis of EDRF and the cerebroarterial response to vasodilator nerve stimulation. Life Sci 47:345–351

Toda N, Ayajiki K, Yoshida K, Kimura H, Okamura T (1993) Impairment by damage of the pterygopalatine ganglion of nitroxidergic vasodilator nerve function in canine cerebral and retinal arteries. Circ Res 72:206–213

Toda N, Kitamura Y, Okamura T (1994) Role of nitroxidergic nerve in dog retinal arterioles in vivo and arteries in vitro. Am J Physiol 266:H1446–H1450

Toda N, Zhang JX, Ayajiki K, Okamura T (1995a) Mechanisms underlying endothelium-independent relaxation by acetylcholine in canine retinal and cerebral arteries. J Pharmacol Exp Ther 274:1507–1512

Toda N, Kitamura Y, Okamura T (1995b) Functional role of nerve-derived nitric oxide in isolated dog ophthalmic arteries. Invest Ophthalmol Vis Sci 36:563–570

Toda N, Ayajiki K, Okamura T (1996) Inhibition of nitroxidergic nerve function by neurogenic acetylcholine in monkey cerebral arteries. J Physiol

Wang Y, Okamura T, Toda N (1993) Mechanisms of acetylcholine-induced relaxation in dog external and internal ophthalmic arteries. Exp Eye Res 57:275–281

Wiencke AK, Nilsson H, Nielsen PJ, Nyborg NCB (1994) Nonadrenergic noncholinergic vasodilation in bovine ciliary artery involves CGRP and neurogenic nitric oxide. Invest Ophthalmol Vis Sci 35:3268–3277

Yoo K, Tschudi, Flammer J, Lüscher TF (1991) Endothelium-dependent regulation of vascular tone of the porcine ophthalmic artery. Invest Ophthalmol Vis Sci 32: 1791–1798.

IV. The Uvea: Pathophysiology of Uveitis

Nitric Oxide in the Iris Sphincter Muscle

HIDEKI CHUMAN and TOMOMI CHUMAN

Introduction

The aim of this chapter is to describe the effect of nitric oxide (NO) on the iris sphincter muscle. NO is a labile free radical, and it has a role as a neurotransmitter as well (Moncada et al. 1991; Grozdanovic et al. 1994). Studies have demonstrated that NO relaxes various kinds of smooth muscles, including those of the respiratory (Sekizawa et al. 1993), digestive (Stark and Szurszewski 1992; Wiklund et al. 1993), genitourinary (Persson and Andersson 1992; Ehren et al. 1994), and vascular (Moncada et al. 1991) systems. There are few reports in terms of the effect of NO on the iris sphincter muscle. We have already clarified the effect of NO on the rabbit iris sphincter muscle for the first time (Chuman et al. 1996, 1997); here we summarize the effect of NO on the iris sphincter muscle (Fig. 1). We noted in rabbits that, the increased accumulation of cyclic guanosine monophosphate (cGMP) induced by sodium nitroprusside (SNP) (an NO donor) in the iris sphincter muscle inhibits cholinergic muscular contraction but does not affect tachykinergic muscular contraction. These results suggest that the different effects on cGMP are essential for the different responses to NO in cholinergic and tachykinergic muscular contractions. Furthermore, this NO–cGMP pathway is operative in vivo for the modulation of iris sphincter muscle tone, at least when the eyes are infected with bacteria.

Anatomy of Iris Sphincter Muscle and Localization of NO Synthase

The iris sphincter muscle is 0.5–1.0mm wide and 40–80μm thick. It consists of bundles of smooth muscle fibers that encircle the pupil in the posterior stroma of the iris, closely approaching the pupillary margin. The vascular arcades from

Department of Ophthalmology, Miyazaki Medical College, 5200 Kihara Kiyotake, Miyazaki 889-16, Japan

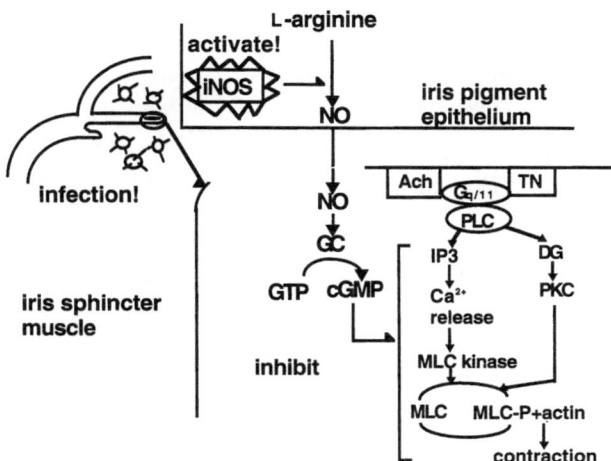

FIG. 1. Schema of NO in the iris sphincter muscle. *iNOS*, inducible nitric oxide synthase; *NO*, nitric oxide; *GC*, guanylate cyclase; *GTP*, guanosine triphosphate; *cGMP*, cyclic guanosine monophosphate; *PLC*, phospholipase C; *IP₃*, inositol trisphosphate; *DG*, diacylglycerol; *MLC*, myosin light chain; *PKC*, protein kinase C

the minor circle extend toward the pupil and through the sphincter muscle (Fig. 2). The pigment epithelium is just posterior to the iris sphincter muscle (Fig. 2). When considering the formation of NO in the iris, we must recognize the site of NO synthase (NOS). Osborne et al. (1993) demonstrated the presence of NOS activity in the rabbit iris epithelial cells by histochemical staining using nicotinamide adenine dinucleotide phosphate (NADPH) diaphorase (Fig. 3). They reported that NOS was generally free in the rabbit iris sphincter muscle. Yamamoto et al. (1993) also suggested that few immunoreactive fibers were encountered in the iris of the rat.

Expression of inducible NOS in the epithelial cells of the iris from endotoxin-induced uveitis was demonstrated (Jacquemin et al. 1996). NO is a gas and can be diffused at 50 μm per second. The studies of Jacquemin et al. suggested that NO may play a role in regulating the contractility of these muscles particularly in the presence of inflammation. We investigated the effect of exogenous NO on contractions of the rabbit iris sphincter muscle.

Rabbit Iris Sphincter Muscle: Physiological and Pharmacological Properties

The iris sphincter muscle, a bundle of circularly arranged smooth muscle cells, is innervated by cholinergic nerves, the activation of which induces muscular contraction. Tachykinin-like contractile responses to sensory nerve stimuli have also

FIG. 2. Pupillary portion of the iris. The sphincter muscle (*c*) and vascular arcade (*d*) are seen. The sphincter muscle and the iris epithelium (*j*) are close to each other at the pupillary margin. (From Hogan et al., 1971, with permission)

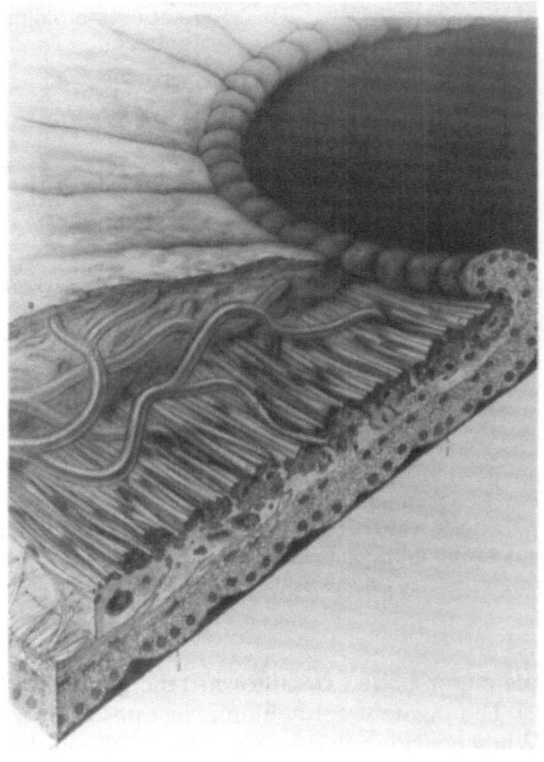

FIG. 3. Section showing the distrbution of NADPH diaphorase in the iris. Intense staining (*arrows*) of NADPH diaphorase is associated with the epithelial cell. (From Osborne et al., 1993, with permission)

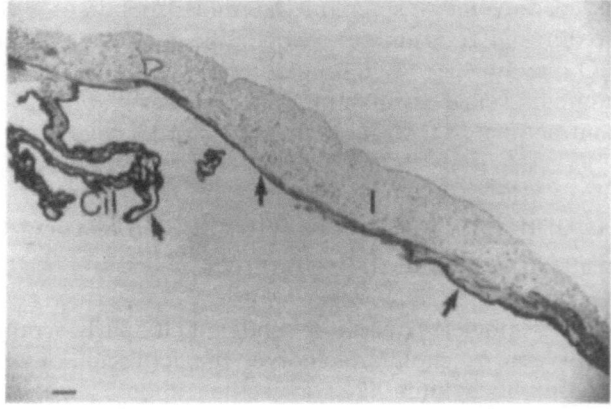

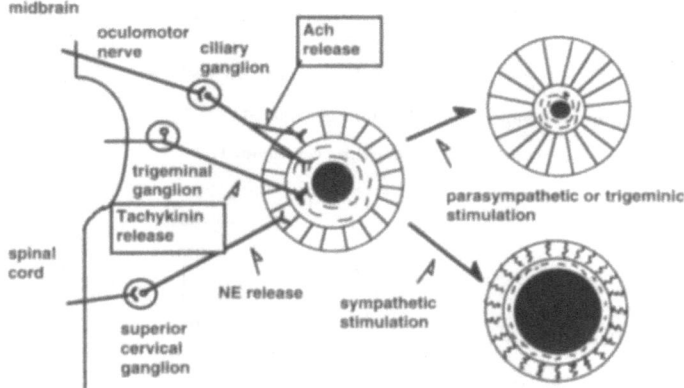

FIG. 4. Simple schema of the pupillary constriction-dilating system under autonomic nerve innervation. Acetylcholine (*Ach*) liberated from the parasympathetic nerve ending produces miosis. Norepinephrine liberated from the sympathetic nerve ending produces mydriasis. In the rabbit, tachykinin liberated from the trigeminal nerve ending produces miosis

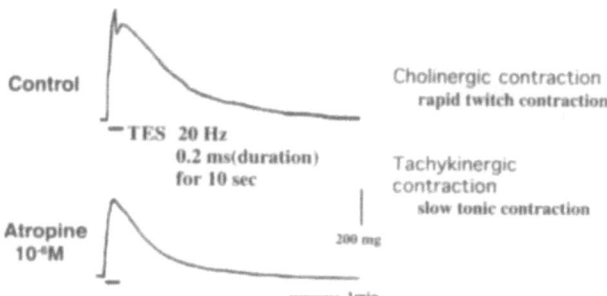

FIG. 5. Transmural electrical stimulation (TES) of the rabbit iris sphincter muscle produces two contractile responses: The rapid twitch contraction is inhibited by atropine, so it is known to be a cholinergic contraction. The slow tonic contraction is inhibited by capsicin, so it is known to be a tachykinergic contraction

been reported in the rabbit iris sphincter muscle (Ueda et al. 1981) (Fig. 4). Transmural electrical stimulation (TES) of the rabbit iris sphincter muscle induces a two-phrase contraction (Fig. 5). The rapid twitch contractions are inhibited by atropine and so are known to be cholinergic contractions. Slow tonic contractions are inhibited by capsaicin, so they are recognized as tachykinergic contractions. We used carbachol (10^{-6} mol/l) and neurokinin A (10^{-7} mol/l) as contractile agonists (Fig. 6; Table 1).

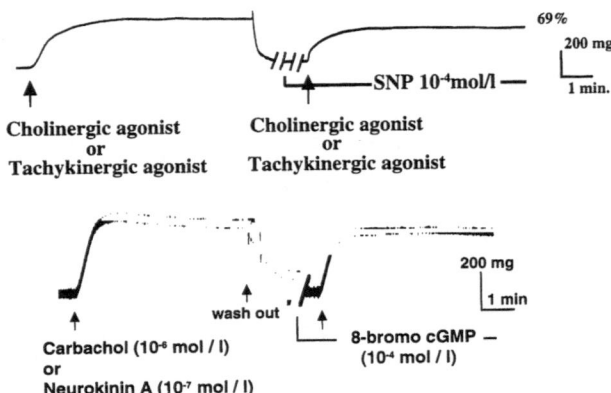

Fig. 6. Schema of the experiment on motor activity of the iris sphincter muscle. After obtaining the maximal contraction following the first administration of carbachol or neurokinin A, the preparation was washed with fresh Krebs' solution twice and gradually was relaxed to a steady-state level. The second administration of carbachol or neurokinin A was then undertaken. All isometric force measurements are given as relative values in comparison with the maximal contraction induced by the first administration of carbachol or neurokinin A. Thus effects of pretreatment with sodium nitroprusside (SNP), carboxy-2-phenyl-4,4,5,5-tetramethyl-imidazoline-1-oxyl-3-oxide (C-PTIO), and 8-bromo cGMP on the second contractions were expressed as percentages of the first maximal contraction caused by carbachol or neurokinin A. When SNP, C-PTIO, and 8-bromo cGMP were used, preparations were treated 10 min before the second administration of carbachol or neurokinin A

TABLE 1. Pharmacological agents

Agonists
　　Cholinergic agonist: carbachol (10^{-6} mol/l)
　　Tachykinergic agonist: neurokinin A (10^{-7} mol/l)
NO-related agents
　　NO donor: sodium nitroprusside (SNP) (10^{-6}, 3×10^{-6}, 10^{-5}, 10^{-4} mol/l)
　　NO scavenger: carboxy-PTIO (C-PTIO) (10^{-4} mol/l, 5×10^{-4} mol/l)
　　NO synthase inhibitor: N^{G}-nitro-L-arginine methyl ester (L-NAME) (10^{-4} mol/l)

Contraction of Iris Sphincter Muscle: Intracellular Mechanism

The intracellular mechanism of the contraction of the iris sphincter muscle in the agonist–receptor system is shown in Fig. 7. Agonists binding to the receptors activate G-protein and stimulate phospholipase C (PLC) to hydrolyze phosphatidylinositol 4,5-bisphosphate (PIP_2) into inositol trisphosphate (IP_3) and diacylglycerol (DG). IP_3 binding to the IP_3 receptor on the sarcoplasmic reticulum

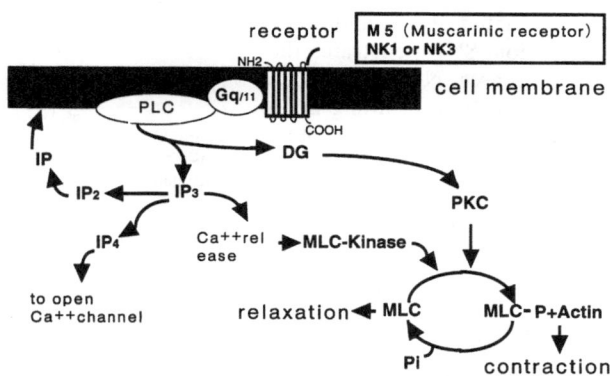

Fɪɢ. 7. Intracellular mechanism of the contraction of iris sphincter muscle. Agonists bind to the receptors to activate the G-protein and stimulate phospholipase C (*PLC*) to inositol trisphosphate (*IP₃*) and diacylglycerol (*DG*). IP₃ releases Ca²⁺. Ca-calmodulin activates a myosin light chain (*MLC*) kinase to phosphorylate a MLC and create a contraction. DG acts by binding to protein kinase C (*PKC*). PKC phosphorylates an MLC and creates a contraction

(SR) releases Ca^{2+}. Ca^{2+} binding to calmodulin (Ca^{2+}–calmodulin activates myosin light chain (MLC) kinase to phosphorylate MLC and induces contraction. IP₃ is then metabolized by a series of dephosphorylation and phosphorylation pathways. DG acts by binding to protein kinase C (PKC). PKC also phosphorylates MLC and induces contraction.

Effect of NO on Rabbit Iris Sphincter Muscle: Pharmacological Approach

To clarify the involvement of NO in the postsynaptic regulation of the iris sphincter muscle tone in rabbits, we examined the effects of SNP as NO donor, carboxy-2-phenyl-4,4,5,5-tetramethyl-imidazoline-1-oxyl-3-oxide (C-PTIO) as NO scavenger, and N^G-monomethyl-ʟ-arginine (ʟ-NAME) as inhibitor of NO synthase on carbamylcholine (carbachol)-induced cholinergic contraction and neurokinin A-induced tachykinergic contraction in the isolated iris sphincter muscle (Figs. 6, 8; Table 1). These pharmacological agents are widely accepted for the investigation of NO.

Rabbit Iris Sphincter Muscle: Motor Activity

Sodium nitroprusside (NO donor) significantly decreased the muscular contraction caused by carbachol in a concentration-dependent manner, whereas the muscular contraction caused by neurokinin A was not affected by SNP even at high concentrations (Figs. 6, 9). In the presence of C-PTIO at 10^{-4} mol/l, the

FIG. 8. Schema of NO-related agents and intracellular signal transduction. SNP penetrates the
cell membrane and releases NO to activate
soluble guanylate cyclase (*sGC*), resulting in an
accumulation of cGMP and a series of physiological responses

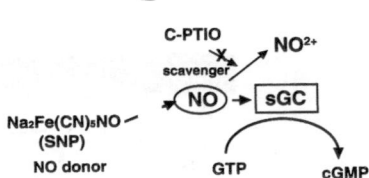

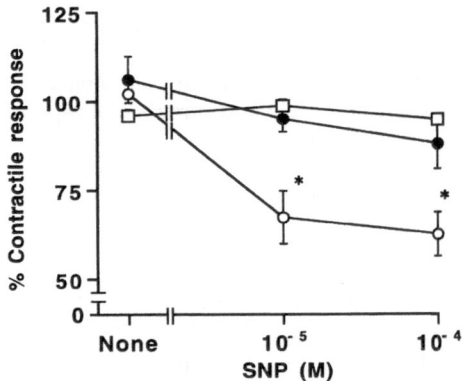

FIG. 9. Effects of sodium nitroprusside (*SNP*) on carbachol- and neurokinin A-induced
contractions of the isolated rabbit iris sphincter muscle in the absence and presence of
C-PTIO. Various concentrations of SNP (10^{-5} to 10^{-4} mol/l), C-PTIO (10^{-4} mol/l), or both
were added to the organ bath between the first and second administrations of carbachol
(10^{-6} mol/l) or neurokinin A (10^{-7} mol/l). Results are means ± SEM ($n = 5$). The value of
100% corresponds to the contraction caused by the first administration of carbachol and
neurokinin A. *Open circles*, carbachol; *closed circles*, C-PTIO + carbachol; *open squares*,
neurokinin A. *$P < 0.05$ indicates a significant difference from the contraction caused by
the second administration of carbachol without SNP (*None*)

inhibitory effects of SNP at 10^{-5} and 10^{-4} mol/l on carbachol-induced muscular
contraction were significantly diminished from 67% to 95% and from 63% to
88%, respectively (Fig. 10). These findings confirm that cholinergic contraction
of the iris sphincter muscle is NO-sensitive, whereas tachykinergic contraction
of that muscle is NO-insensitive.

Intracellular Signal Transduction by NO

Intracellular signal transduction by NO is thought to be triggered by activation
of a soluble guanylate cyclase, resulting in accumulation of cGMP and a series
of physiological responses (Ignaro 1991; Moncada and Higgs 1993) (Fig. 7). Some

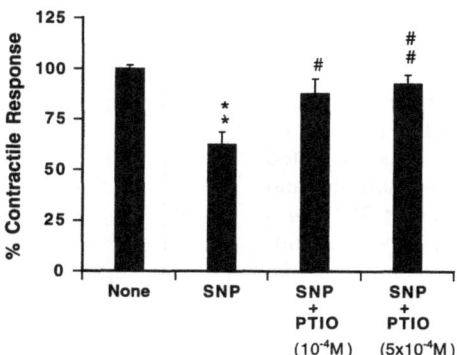

Fig. 10. Effect of C-PTIO on the inhibition of carbachol-induced contraction by SNP. The preparation was treated with SNP (10^{-4} mol/l), C-PTIO (10^{-4} or 5×10^{-4} mol/l), or both between the first and second administrations of carbachol (10^{-6} mol/l). Results are means ± SEM ($n = 7$). The value of 100% corresponds to the contraction induced by the first administration of carbachol. **$P < 0.0005$ indicates a significant difference from the contraction induced by the second administration of carbachol without SNP (*None*). $^{#}P < 0.01$ and $^{##}P < 0.005$ indicate significant differences from the contraction induced by the second administration of carbachol with SNP alone

studies, however, indicate that NO participates in other types of molecular mechanisms. For example, Smith and Li (1993) reported a novel effect of NO as mediator of N-methyl-D-aspartate (NMDA)-induced phosphatidylinositol hydrolysis in the neonatal rat cerebellum. In contrast, Bolotina et al. (1994) reported that NO directly activates calcium-dependent potassium channels in the vascular smooth muscle. In our experiments (Chuman et al. 1996), methylene blue, an inhibitor of soluble guanylate cyclase, significantly diminished the inhibitory effect of SNP on cholinergic contraction. In addition, SNP alone increased accumulation of cGMP in the muscle in a concentration-dependent manner. Thus, our findings indicate that NO suppresses cholinergic contraction of the iris sphincter muscle mainly through a cGMP-dependent mechanism. To clarify the mechanism(s) for the different responses to NO in cholinergic and tachykinergic contractions of rabbit iris sphincter muscle, we measured the accumulation of cGMP in the muscle after treatment of NO-related agents and examined the effects of NO-related agents and 8-bromo cGMP on the motor activity of the muscle caused by carbachol or neurokinin A in vitro.

Effect of SNP on cGMP Content in Iris Sphincter Muscle

The cGMP content in the muscle before drug treatment was 1.28 ± 0.09 pmol/mg protein ($n = 5$). SNP at 10^{-6}, 10^{-5}, and 10^{-4} mol/l increased the accumulation of cGMP in the muscle in a concentration-dependent manner to 206%, 670%,

Fig. 11. Effect of SNP on the cGMP content in the isolated rabbit iris sphincter muscle. The preparation was treated without or with SNP in various concentrations (10^{-6} to 10^{-4} mol/l) for 10 min. After incubation, cGMP was extracted and assayed as described in Materials and Methods. Results are means ± SEM ($n = 5$). ***$P < 0.0001$ indicates a significant difference from the cGMP content in the tissue without SNP treatment (*None*)

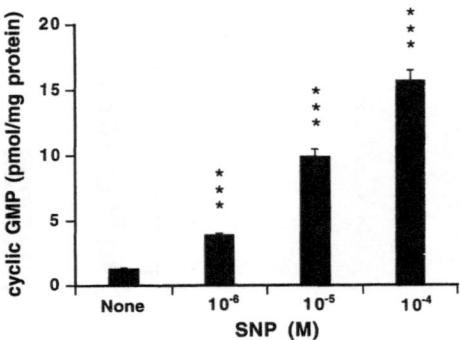

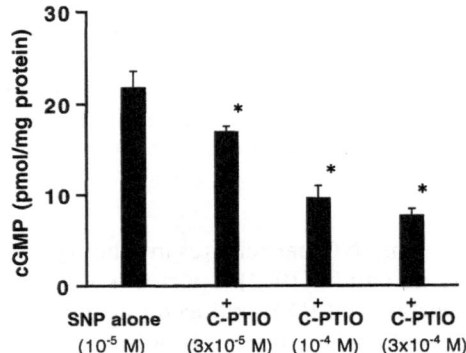

FIG. 12. Effect of carboxy-PTIO (*C-PTIO*) on sodium nitroprusside (*SNP*)-induced accumulation of cGMP in the isolated rabbit iris sphincter muscle. The preparation was treated with or without SNP (10^{-5} mol/l) in combination with or without C-PTIO in various concentrations (3×10^{-5} to 3×10^{-4} mol/l) for 10 min. After incubation, cGMP was extracted and assayed as when iNOS, induced by ocular infection with bacteria, catalyzes excess formation of NO. The NO–cyclic GMP pathway appeared to be pathophysiologically important for regulating the cholinergic contractility of the iris sphincter muscle tone

and 1200% ($n = 5$), respectively (Fig. 11). These findings suggest that SNP stimulates cGMP formation, which in turn inhibits carbachol stimulation of muscle contraction. To establish that this effect of SNP is related to the NO liberated from this drug, we observed the effect of C-PTIO on SNP-induced cGMP accumulation in the muscle. C-PTIO decreased, in a concentration-dependent manner, the cGMP accumulation induced by SNP (10^{-5} mol/l) to 78%, 44%, and 35% at 3×10^{-5}, 3×10^{-4}, and 3×10^{-4} mol/l, respectively (Fig. 12). These results indicate that NO liberated from SNP increases the cGMP accumulation in the muscle.

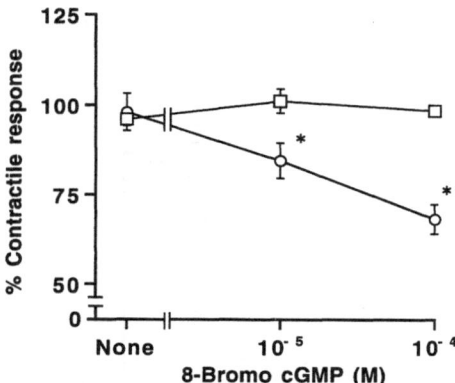

FIG. 13. Effect of 8-bromo cyclic guanosine monophosphate (8-bromo cGMP) on carbachol- and neurokinin A-induced contractions of the isolated rabbit iris sphincter muscle. Various concentrations of 8-bromo cGMP (10^{-5}–10^{-4} mol/l) were added to the organ bath between the first and second administrations of carbachol (10^{-6} mol/l) or neurokinin A (10^{-7} mol/l). Values shown are means ± SEM ($n=5$). A value of 100% corresponds to the contraction induced by the first administration of carbachol and neurokinin A. *Circles*, carbachol; *squares*, neurokinin A. *Asterisks*, $P<0.05$, indicate significant difference from the contraction caused by the second administration of carbachol without 8-bromo cGMP (*none*)

Does NO Act on Iris Sphincter Muscle Through Formation of cGMP?

Although we observed that SNP increased the accumulation of cGMP in rabbit iris sphincter muscle, there was no direct evidence linking the increased accumulation of cGMP induced by SNP to the inhibitory effect of SNP on cholinergic contraction of that muscle. To examine the effects of increased cGMP accumulation in the iris sphincter muscle on cholinergic and tachykinergic muscular contractions, we treated the muscle with 8-bromo cGMP between the first and second administrations of carbachol (10^{-6} mol/l) or neurokinin A (10^{-7} mol/l) (Fig. 6). As shown in Fig. 13, 8-bromo cGMP at 10^{-5} and 10^{-4} mol/l significantly inhibited the carbachol-induced muscular contraction to 86% and 70%, respectively, but had no effect on neurokinin A-induced muscular contraction. These results show that the increased accumulation of cGMP in the iris sphincter muscle is able to induce depression of cholinergic muscular contraction but has no effect on tachykinergic muscular contraction.

NO–cGMP Pathway

To determine whether the intracellular signal transduction pathway stimulated by carbachol or neurokinin A influences SNP-induced NO formation, the content of cGMP in the muscle was measured after treatment with carbachol

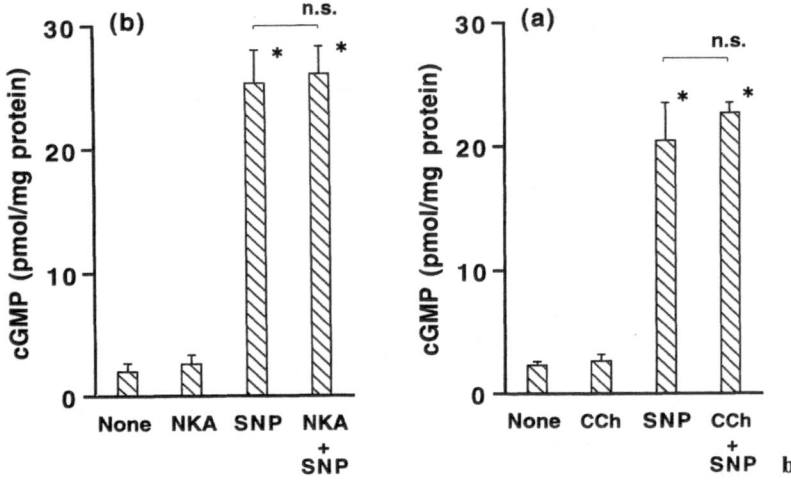

FIG. 14. Effects of carbachol (**a**) and neurokinin A (**b**) on sodium nitroprusside (*SNP*)-induced accumulation of cGMP in the isolated rabbit iris sphincter muscle. The preparation was treated with or without carbachol (*CCh*, 10^{-6} mol/l), neurokinin A (*NKA*, 10^{-7} mol/l), or sodium nitroprusside (*SNP*, 10^{-5} mol/l) for 10 min. The results are means ± SEM ($n = 4$). *$P < 0.05$ indicates a significant difference from the cGMP accumulation in the muscle treated without drug treatment (*None*). *n.s.*, nonsignificant difference from the cGMP accumulation by SNP alone

(10^{-6} mol/l) or neurokinin A (10^{-7} mol/l) in the presence or absence of SNP (10^{-5} mol/l). Neither carbachol (Fig. 14a) nor neurokinin A (Fig. 14b) influenced the basal cGMP content and the SNP-induced accumulation of cGMP in the muscle, suggesting that carbachol and neurokinin A have no properties as a scavenger of NO radicals or a superoxide dismutate, which enhances the stability of NO radicals.

No-insensitive Tachykinergic Contraction of Iris Sphincter Muscle

Sodium nitroprusside failed to suppress tachykinergic contraction of the iris sphincter muscle. There are some possible explanations for this failure. The first possibility is that the tachykinergic stimulus causes excess formation of endogenous NO; therefore subsequent administration of exogenous NO no longer inhibits the tachykinergic muscular contraction. This idea, however, has been negated by our study in which L-NAME, an inhibitor of NO formation derived from L-arginine, did not affect the tachykinergic muscular contraction (Chuman et al. 1996). The second and third possibilities are that the tachykinergic stimulus prevents the NO-induced accumulation of cGMP in the muscle by inhibiting soluble guanylate cyclase and inducting the NO radical scavenger, respectively.

As mentioned above, neurokinin A and carbachol did not affect the basal cGMP content or the SNP-induced increased cGMP content in the muscle. Therefore, these explanations for the tachykinergic response are less likely. Our results ultimately suggested the fourth possibility: that the intracellular signal transduction pathway for excitation-contraction coupling in the tachykinergic response is not influenced by the increased cGMP accumulation in the muscle, whereas that of the cholinergic response is suppressed. In fact, we observed that 8-bromo cGMP, a permeable cGMP analog, significantly diminished the muscular contraction caused by carbachol but not that due to neurokinin A. These results suggest that the different sensitivities of cGMP are essential for the different responsiveness to NO in cholinergic and tachykinergic muscular contractions. It is possible that cholinergic and tachykinergic responses have a distinct signaling pathway downstream from PIP_2 hydrolysis that plays a role in the excitation-contraction coupling in the rabbit iris sphincter muscle (Fig. 8): Carbachol and neurokinin A induce activation of phospholipase C-β through a guanine nucleotide-binding regulatory protein (G-protein) (Wess et al. 1993; Regoli et al. 1994) (Tables 2, 3) termed $G_{q/11}$ and therefore stimulates PIP_2 hydrolysis (Abdel-Latif 1989; Tachado et al. 1991; Taniguchi et al. 1992). Thus, the former cholinergic pathway appears to be more sensitive than the latter tachykinergic pathway upon cGMP.

Does the NO–cGMP Pathway Work In Vivo?

Our next objective was to determine whether the NO–cGMP pathway is operative in vivo for regulating iris sphincter muscle tone. Several isoforms of NOS have been reported and subclassified into two categories: constitutive NOS (cNOS) and inducible NOS (iNOS). cNOS is constitutively expressed in endothelial cells, forebrain, platelets, noradrenergic noncholinergic (NANC) nerves, and adrenal medullary cells for the physiological control of a variety of functions (Moncada et al. 1991; Mandai et al. 1994). iNOS is induced only when macrophages, Kupffer cells, hepatocytes, vascular smooth muscle cells, and fibroblasts are stimulated by various signals, such as lipopolysaccharide (LPS) or cytokines, and immediately catalyzes excess formation of NO (Moncada et al. 1991). Incidentally, SNP, employed as a NO donor in our experiments, liberates NO from this drug without the activity of NOS (Ahlner et al. 1991). Mandai et al. (1994) studied the changes in NOS activity associated with LPS-elicited ocular inflammation (endotoxin-induced uveitis) in rats. They reported that the induction of iNOS may play a key role in the ocular inflammation elicited by LPS because N^G-nitro-L-arginine, an inhibitor of NO synthesis, reduced the inflammatory response. In our experimental system, it is difficult to investigate cNOS-induced NO formation. Thus, we measured accumulation of cGMP in the iris sphincter muscle dissected from rabbit eye pretreated with intravitreal injection of LPS, a stimulator of iNOS induction in vivo.

Table 2. Acetylcholine receptors (muscarinic)

Nomenclature[a]	M1	M2	M3	M4	m5
Antagonists[b]	MT7 (9.8)	Tripitramine (9.4–9.6)	4-DAMP (8.9–9.3)	MT3 (8.7)	4-DAMP (8.9–9.0)
	4-DAMP (8.6–9.2)	AFDX384 (8.2–9.0)	Darifenacin (8.4–8.9)	4-DAMP (8.4–9.4)	Darifenacin (8.0–8.1)
	Tripitramine (8.4–8.8)	Himbacine (8.0–8.3)	AFDX384 (7.2–7.8)	Himbacine (8.0–8.8)	Tripitramine (7.3–7.5)
	Pirenzepine (7.8–8.5)	4-DAMP (7.8–8.4)	Tripitramine (7.1–7.4)	AFDX384 (8.0–8.7)	Guanylpirenzepine (6.8)
	Guanylpirenzepine (7.7)	Darifenacin (7.0–7.4)	Himbacine (6.9–7.4)	Tripitramine (7.8–8.2)	Pirenzepine (6.2–7.1)
	Darifenacin (7.5–7.8)	Pirenzepine (6.3–6.7)	Pirenzepine (6.7–7.1)	Darifenacin (7.7–8.0)	AFDX384 (6.3)
	AFDX384 (7.3–7.5)	MT7 (<6)	Guanylpirenzepine (6.5)	PD102807 (7.3)	Himbacine (6.1–6.3)
	MT3 (7.1)	MT3 (<6)	PD102807 (6.2)	Pirenzepine (7.1–8.1)	MT3 (<6)
	Himbacine (7.0–7.2)	PD102807 (5.7)	MT3 (<6)	Guanylpirenzepine (6.5)	MT7 (<6)
	PD102807 (5.3)	Guanylpirenzepine (5.5)	MT7 (<6)	MT7 (<6)	PD102807 (5.2)
Radioligands[c]	[3H]pirenzepine	—	—		—
G-protein effector	Gq/11	Gi/o	Gq/11	Gi/o	Gq/11
Gene/chromosomal localization	CHRM1/11	CHRM2/7q35–36	CHRM3/1q43–44	CHRM4/11p12–11.2	CHRM5/15q26
Structure—7TM	h460 P11229	h466 P08172	h590 P20309	h479 P08173	h532 P08912
	m460 P12657	r466 P10980	r589 P08483	m479 P32211	r531 P08911
	r460 P08482			r478 P08485	

From 1998 Receptor and ion channel nomenclature supplement, 9th ed. Compiled in association with the IUPHAR.
The M5 receptor is in the rabbit iris sphincter muscle. As shown in the table, the effector is Gq1.
[a] Nomenclature proposed by the NC-IUPHAR Subcommittee on Muscarinic Acetylcholine Receptors.
[b] MT3 and MT7 are toxins contained within the venom of the green (Dendroaspis augusticeps) and black (Dendroaspis polylepis) mamba (Jerusalinsky and Harvey, 1994).
[c] [3H](−)-N-methylscopolamine and [3H](−)-3-quinuclidinylbenzilate are nonselective radioligands and have K_d values of less than 1.0 nM at all subtypes.

TABLE 3. Tachykinin receptors

Nomenclature	NK$_1$	NK$_2$	NK$_3$
Other names	Substance P SP-P	Substance K SP-E/SP-K	Neurokinin B SP-E/SP-N
Potency order	SP > NKA > NKB	NKA > NKB ≫ SP	NKB > NKA > SP
Selective agonists	SP methylester [Sar^9,Met(O$_2$)^{11}]SP [Pro^9]SP	[β-Ala^8]NKA-4–10 [Lys^5, MeLeu^9,Mle^{10}]NKA-4–10 GR64349	—
Selective antagonists	SR140333 (9.5[a]; 10.6[b]) LY303870 (9.4) CP99994 (9.3[a]; 7.3[b]) RP67580 (7.6[a]; 8.5[b])	GR94800 (9.6) GR159897 (9.5) SR48968 (9.0) MEN10627 (8.6)	SR142802 (9.2) SB223412 (9.0)[c] PD157672 (7.8)
Radioligands	[^{3}H] or [^{125}I]SP [^{3}H] or [^{125}I]BH- [Sar^9,Met(O$_2$)^{11}]SP [^{125}I]L703606 (0.3 nM)	[^{3}H]SR48968 (0.5 nM) [^{3}H]GR100679 [^{125}I]NKA	[^{3}H]senktide [^{3}H]SR142802 (0.13 nM) [^{125}I][MePhe^7]NKB
G-protein effector Gene/chromosomal localization	G$_{q/11}$ NKR1/2	G$_{q/11}$ NKR2/10q11–21	G$_{q/11}$ NKR3
Structure—7TM	h407 P25103 m407 P30548 r407 P14600	h398 P21452 m384 P30549 r390 P16610	h468 P29371 r452 P16177

From 1998 Receptor and ion channel nomenclature supplement, 9th ed. Compiled in association with the IUPHAR.
NK1 and NK3 receptor is in the rabbit iris sphincter muscle. As shown in the table, the effector is Gq11.
Comment: Marked species differences in pharmacology exist for all three receptors, in particular with non-peptide ligands. The hexapeptide agonist septide appears to bind to an overlapping but nonidentical site to SP on the NK$_1$ receptor.
Other receptors/binding sites: There is evidence from functional and radioligand binding studies to suggest sites for SP-1–7 that fail to recognize other tachykinis.
Endogenous ligand(s): Substance P (SP), neurokinin A (NKA; previously known as substance K, neu-rokinin α, neuromedin L), neurokinin B (NKB; previously known as neurokinin β, neuromedin K), neu-ropeptide K and neuropeptide γ (N-terminally extended forms of neurokinin A); neurokinins are mammalian members of the tachykinin family, which includes peptides of mammalian and nonmammalian origin containing the consensus sequence Phe-X-Gly-Leu-Met.
[a] pK_i at human receptor.
[b] pK_i at rat receptor.
[c] Sarau HM et al. (1997) *J Pharmacol Exp Ther* 281:1303–1311.

After treating the eyes with LPS for 9, 12, 15, and 24h, cGMP accumulation in muscle was increased in a time-dependent manner (Fig. 15). L-NAME (50mg/kg i.p.) had no discernible effect on basal cGMP content in the iris sphinc-ter muscle (data not shown) but significantly reduced the LPS-elicited cGMP accumulation at 15h to 17% (Fig. 15). These results suggest that the NO–cGMP pathway is also operative in the iris sphincter muscle in vivo, at least when iNOS, induced by ocular infection with bacteria, catalyzes excess formation of NO. Finally, the NO–cGMP pathway appeared to be pathophysiologically important for regulating cholinergic contractility of the iris sphincter muscle tone.

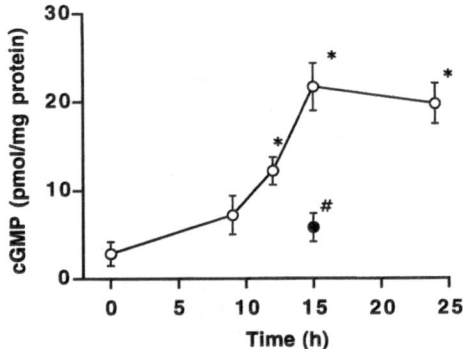

FIG. 15. Effects of the intravitreal injection of lipopolysaccharide (LPS) on the content of cGMP in the rabbit iris sphincter muscle with or without pretreatment of N^G-nitro-L-arginine methyl ester (L-NAME) in vivo. Rabbits were treated with or without LPS (20 ng/20 µl; intravitreal injection) and L-NAME (50 mg/kg i.p., repeated injections) for the indicated period as described in Materials and Methods, and then the iris sphincter muscle was dissected. Results are means ± SEM (n = 4). *Open circles*, LPS; *closed circle*, L-NAME + LPS. *P < 0.05 indicates a significant difference from the basal cGMP content in the tissue at 0 h. #P < 0.05 indicates a significant difference from the cGMP accumulation elicited by LPS alone at 15 h

Other Reports of NO Effects on Iris Sphincter Muscle

There are few reports about the effect of NO on iris sphincter muscle other than ours. Ding and Abdel-Latif (1997) suggested that SNP attenuated carbachol-induced contraction and IP$_3$ production through the formation of cGMP in the rabbit iris sphincter muscle. These effects are tissue- and species-specific. Their results are in agreement with ours. Wiklund et al. (1993) observed that complex contraction by transmural electrical stimulation (TES) was influenced by neither acid nitrate (an NO donor) nor L-NAME; and they speculated that NO does not modulate cholinergic or tachykinergic neurotransmission. Our finding that contraction of the iris sphincter muscle induced by carbachol is NO-sensitive differs from a previous one obtained for contraction caused by TES. It has been reported that sensitivity to the relaxant effect of NO donors varies considerably among smooth muscle preparations and depends on the experimental conditions used (Ahlner et al. 1991). Therefore, a possible explanation for the difference between the two studies may be the use of different NO donors. Another possibility is that we evaluated only the postsynaptic effect of NO on individual contractions induced by cholinergic or tachykinergic agonists, whereas in the other study the pre- and postsynaptic effects of NO on TES-induced contraction were evaluated simultaneously. In the association with prostaglandin (PG), Astin et al. (1994) demonstrated that PGF$_{2\alpha}$ causes surface hyperemia of the eye by activating NOS, but this mechanism seems to be partly involved in the increased blood flow of the iris. Yousufzai et al. (1997) observed that SNP (100 M) increased the release of $PGF_{2\alpha}$ and PGE_2 by 64% and 70%, respectively, in dog iris sphincter muscle. There seem to be no reports about an association between NO and endothelin in the iris sphincter muscle.

Conclusion and the Future

Nitric oxide inhibits cholinergic contraction but not tachykinergic contraction through the formation of cGMP in the rabbit iris sphincter muscle. Because contraction of the human iris sphincter muscle is innervated only by cholinergic nerves, the inhibitory role of NO in regulating the contractility of this muscle may be more important in humans than in rabbits. Clarification of the intracellular mechanisms of cholinergic and tachykinergic responses is of great interest.

Once the eye is infected with bacteria, NO produced from iNOS helps to kill bacteria and possibly prevents miosis. It may inhibit miosis owing to the induction of prostaglandins. Furthermore, we have found an increase in NO level in aqueous humor after intraocular surgery and have suggested that NO can affect the pupil size during and after intraocular surgery.

Nitric oxide is known to have a short half-life. In the future it may be the ideal mydriatic eyedrop, enlarging the pupil in a short time, with rapid reversion to normal size. NO may have a role in such diseases as tonic pupil. This interesting mysterious gas may help us accumulate information about treating and curing disease. We look forward to future studies on NO.

References

Abdel-Latif AA (1989) Calcium-mobilizing receptors, polyphosphoinositides, generation of second messengers and contraction in the mammalian iris smooth muscle: historical perspectives and current status. Life Sci 45:757–786

Abdel-Latif AA (1997) Iris-ciliary body, aqueous humor and trabecular meshwork. In: Harding JJ (ed) Biochemistry of the eye. Chapman & Hall, London, pp 79

Ahlner J, Andersson RGG, Torfgard K, Axelsson KL (1991) Organic nitrate esters: clinical use and mechanisms of actions. Pharmacol Rev 43:351–423

Astin M, Stjernschntz J, Selen G (1994) Role of nitric oxide in PGF_{2alpha}-induced ocular hyperemia. Exp Eye Res 59:401–407

Bolotina VM, Najibi S, Palacino JJ, Pagano PJ, Cohen RA (1994) Nitric oxide directly activates calcium-dependent potassium channels in vascular smooth muscle. Nature 368:850–853

Chuman H, Chuman T, Nao-i N, Sawada A, Yamamoto R, Kobayashi H, Wada A (1997) Different responsiveness to nitric oxide–cyclic guanosine monophosphate pathway in cholinergic and tachykinergic contractions of the rabbit iris sphincter muscle. Invest Ophthalmol Vis Sci 38:1719–1725

Chuman T, Chuman H, Nao-i N, Sawada A, Yamamoto R, Wada A (1996) Nitric oxide-sensitive and -insensitive contractions of the isolated rabbit iris sphincter muscle. Invest Ophthalmol Vis Sci 37:1437–1443

Ding KH. Abdel-Latif AA (1997) Actions of C-type natriuretic peptide and sodium nitroprusside on carbachol-stimulated inositol phosphate formation and contraction in ciliary and iris sphincter smooth muscles. Invest Ophthalmol Vis Sci 38:2629–2638

Ehren I, Iversen H, Jansson O, Adolfsson J, Wiklund NP (1994) Localization of nitric oxide synthase activity in the human lower urinary tract and its correlation with neuroeffector responses. Urology 44:683–687

Grozdanovic Z, Brüning G, Baumgarten HG (1994) Nitric oxide: a novel autonomic neurotransmitter. Acta Anat (Basel) 150:16–24

Hogan MJ, Alvalado JA, Weddel JE (1971) Histology of the human eye. Philadelphia, Saunders

Hope BT, Michael GJ, Knigge KM, Vincent SR (1991) Neuronal NADPH diaphorase is a nitric oxide synthase. Proc Natl Acad Sci USA 88:2811–2814

Ignaro LJ (1991) Signal transduction mechanisms involving nitric oxide. Biochem Pharmacol 41:485–490

Jacquemin E, Kozak YD, Thillaye B, Courtois Y, Goureau O (1996) Expression of inducible nitric oxide synthase in the eye from endotoxin-induced uveitis in rats. Invest Ophthalmol Vis Sci 37:1187–1196

Jerusalinsky D, Harvey AL (1994) Toxins from mamba venoms: small proteins with selectivities for different subtypes of muscarinic acetylcholine receptors. Trends Pharmacol Sci 15 (11):424–430

Mandai M, Yoshimura N, Yoshida M, Iwaki M, Honda Y (1994) The role of nitric oxide synthase in endotoxin-induced uveitis: effects of N^G-nitro-L-arginine. Invest Ophthalmol Vis Sci 35:3673–3680

Moncada S, Higgs A (1993) The L-arginine–nitric oxide pathway. N Engl J Med 329:2002–2012

Moncada S, Palmer RMJ, Higgs EA (1991) Nitric oxide: physiology, pathophysiology, and pharmacology. Pharmacol Rev 43:109–142

Osborne NN, Barnett NL, Herrera AJ (1993) NADPH diaphorase localization and nitric oxide synthetase activity in the retina and anterior uvea of the rabbit eye. Brain Res 610:194–198

Persson K, Andersson KE (1992) Nitric oxide and relaxation of pig lower urinary tract. Br J Pharmacol 106:416–422

Regoli D, Boudon A, Fauchére JL (1994) Receptors and antagonists for substance P and related peptides Pharmacol Rev 46:551–599

Sekizawa K, Fukushima T, Ikarashi Y, Maruyama Y, Sasaki H (1993) The role of nitric oxide in cholinergic neurotransmission in rat trachea. Br J Pharmacol 110:816–820

Smith SS, Li J (1993) Novel action of nitric oxide as mediator of N-methyl-D-aspartate-induced phosphatidylinositol hydrolysis in neonatal rat cerebellum. Mol Pharmacol 43:1–5

Stark ME, Szurszewski JH (1992) Role of nitric oxide in gastrointestinal and hepatic function and disease. Gastroenterology 103:1928–1949

Tachado SD, Akhtar RA, Yousufzai SYK, Abdel-Latif AA (1991) Species differences in the effects of substance P on inositol trisphosphate accumulation and cyclic AMP formation, and on contraction in isolated iris sphincter of the mammalian eye: differences in receptor density. Exp Eye Res 53:729–739

Taniguchi T, Ninomiya H, Fukunaga R, Ebii K, Yamamoto M, Fujiwara M (1992) Neurokinin A-stimulated phosphoinositide breakdown in rabbit iris sphincter muscle. Jpn J Pharmacol 59:213–220

Ueda N, Muramatsu I, Sakakibara Y, Fujiwara M (1981) Noncholinergic, nonadrenergic contraction and substance P in rabbit iris sphincter muscle. Jpn J Pharmacol 31:1071–1079

Wess J (1993) Molecular basis of muscarinic acetylcholine receptor function. Trends Pharmacol Sci 14:308–313

Wiklund CU, Olgart C, Wiklund NP, Gustafsson LE (1993) Modulation of cholinergic and substance P-like neurotransmission by nitric oxide in the guinea-pig ileum. Br J Pharmacol 110:833–839

Wiklund CU, Wiklund NP, Gustafsson LE (1993) Modulation of neuroeffector transmission by endogenous nitric oxide: a role for acetylcholine receptor-activated nitric oxide formation, as indicated by measurements of nitric oxide/nitrite release. Eur J Pharmacol 240:235–242

Yamamoto R, Bredt DS, Snyder SH, Stone RA (1993) The localization of nitric oxide synthase in the rat eye and related cranial ganglia. Neuroscience 54:189–200

Yousufzai SY, Abdel-latif AA (1997) Nitric oxide-prostaglandin interactions. Biochemistry of the eye 78–79

Role of Peroxynitrite in Photoreceptor Damage in Experimental Uveitis

GUEY-SHUANG WU and NARSING A. RAO

Introduction

With inflammation, as in uveitis, the phagocytes undergo respiratory burst resulting in the release of a variety of microbicidal oxygen metabolites. The most important primary species of these metabolites are superoxide and hydrogen peroxide, both of which are potentially injurious to surrounding tissues. The superoxide-mediated toxicity in vivo, however, has been difficult to demonstrate because of the low chemical reactivity of the superoxide anion. It was presumed that the cytotoxicity was exerted through formation of highly reactive hydroxyl radicals via the interaction of superoxide with hydrogen peroxide (Weiss and LoBuglio 1982); but the current understanding of cytotoxicity via the reactivity of hydroxyl radicals does not always offer a satisfactory explanation for the experimental results generated (Babbs and Griffin 1989). Therefore, it appears that there must be another pathway to account for the toxicity of superoxide in inflammatory conditions, including the reaction of superoxide with another abundantly generated species, nitric oxide, to form the potent oxidant peroxynitrite. The reaction between superoxide and nitric oxide proceeds at a nearly diffusion-limited rate. The product generated by this reaction is capable of abstracting hydrogen atoms from lipids and nitrating aromatic amino acid residues of cellular macromolecules.

We have demonstrated the presence of both superoxide and nitric oxide in experimentally induced uveitis, commonly known as experimental autoimmune uveitis (EAU) in Lewis rats (Gritz et al. 1991; Zhang et al. 1993). Although the chemistry of peroxynitrite has been elaborated in recent years, the possible role of this oxidant in acute intraocular inflammation has been examined only sparsely (Allen et al. 1996). In this study, we localized both peroxynitrite and its peroxidative products in S-antigen-induced EAU, and we evaluated the possible relation between the lipid peroxide lesions and their likely causative factor, peroxynitrite.

Doheny Eye Institute, 1450 San Pablo Street, Los Angeles, CA 90033, USA

127

Materials and Methods

Experimental autoimmune uveitis was induced in Lewis rats (Charles River Laboratory, Wilmington, MA, USA) by footpad injection of 50 μg of retinal S-antigen as previously described (Rao 1990). Nonimmunized animals were used as controls. The animals were sacrificed at the peak of inflammation, 16 days after immunization. All animal procedures used in this study were conducted in accordance with the ARVO Resolution on the Use of Animals in Research.

Immunofluorescence staining for peroxynitrite was carried out using a rabbit polyclonal anti-nitrotyrosine antibody (Upstate Biotechnology, Lake Placid, NY, USA) as the primary antibody and a goat anti-rabbit immunoglobulin (IgG) conjugated to rhodamine (Jackson Immuno Research Laboratory, West Grove, PA, USA) as the secondary antibody. For localization of lipid peroxide-derived cellular carbonyls, the cryostat sections were reacted with a 0.1% solution of fluorescent 3-hydroxy-2-naphthoic acid (NAH) in saline containing 10% dimethylsulfoxide and 0.095% p-toluenesulfonic acid. The fluorescent emission of the carbonyl-NAH product was collected using a confocal laser scanning microscope. Sections were also coupled with fast blue B (FBB) for color visualization. For in vitro generation of peroxidized retinal lipids, naive Lewis eyes were opened by a circumferential incision, slightly anterior to the ora serrata; and the anterior segment and vitreous were removed. The posterior segments were then incubated aerobically with 200 μM of 2,2′ azobis (2-amidinoproane) hydrochloride (AAPH) (Polysciences, Warrington, PA, USA) in Ringer's buffer for 2 h. Following incubation, the eye cups were sectioned and coupled with NAH and FBB using the same method as the in vivo experiments.

Results

The immunostaining of nitrotyrosine was concentrated in the photoreceptors of the inflamed retina. Most of the staining in this area was of a higher density and was spread to relatively large areas. In the inner retina, positive lesions were also seen in ganglion cell layers, nerve fiber layers, and retinal blood vessels. These inner retinal lesions occur much less frequently than those in the photoreceptors. When the primary antibody was replaced with phosphate-buffered saline, only weak nonspecific staining was seen in the choroid and sclera.

The presence of cellular carbonyls indicative of lipid peroxide formation was detected exclusively in the photoreceptors, and the most intensely fluorescent lesions were concentrated in the photoreceptor outer segments (Fig. 1). The staining in this area was spread to relatively large areas, as would be expected from the chain reaction nature of the lipid peroxidation. No fluorescent lesions were seen in the nonimmunized control animals. The fact that these positive lesions were not localized on the infiltrating polymorphonuclear leukocytes (PMNs) themselves was confirmed by comparing the NAH-FBB staining with

FIG. 1. Confocal imaging of lipid peroxidation products in rat retinas with experimental autoimmune uveitis (EAU). Cryostat sections of posterior segments from inflamed eyes were incubated with fluorescent 3-hydroxy-2-naphthoic acid hydrazide. A Zeiss confocal laser scanning microscope, set at 540 nm for excitation and 590 nm for emission, was used for visualization. Lipid peroxidation products detected as cellular carbonyls were found to localize exclusively in the photoreceptors. (×460)

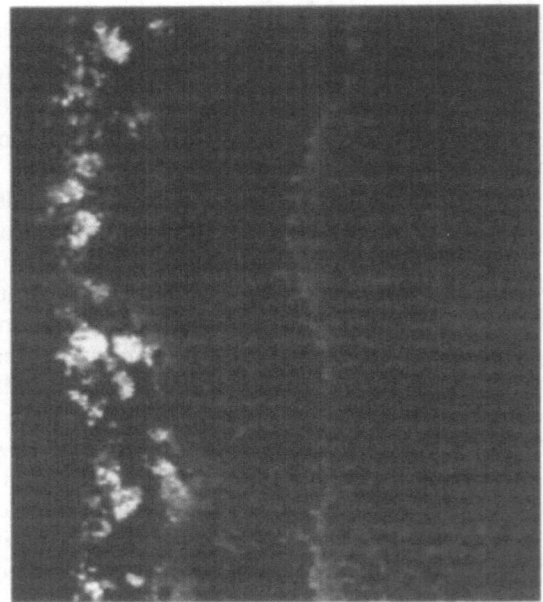

hematocylin-eosin (H-E) staining of the adjacent sections of the inflamed retina.

To ascertain that the NAH staining of the photoreceptors was indeed derived from the lipid peroxidation products, a positive control was generated by incubating the naive Lewis eye cups with a well-known free radical generator, AAPH. In this in vitro peroxidation of retinal lipids, the NAH staining obtained was similar to that seen in the EAU animals in terms of pattern and location (Fig. 2). However, the fluorescent lesions were more intensely stained and more widespread, as would be expected after exposing these lipids to the radical generator in a solution environment.

Discussion

In the present study we demonstrated that there was significant production of peroxynitrite in S-antigen-induced experimental uveitis. The positive staining was concentrated in the photoreceptors, but a small number of lesions were also been in the inner retina, including the nerve fiber layer and inner limiting membranes. The lipid peroxidation products were found to localize exclusively in the photoreceptors, in the same location where peroxynitrite lesions were seen to concentrate. Lipid peroxide lesions, similar to the EAU-originated lesions, were also generated under in vitro conditions, where lipid peroxidation was induced aerobically with the authentic radical generator AAPH.

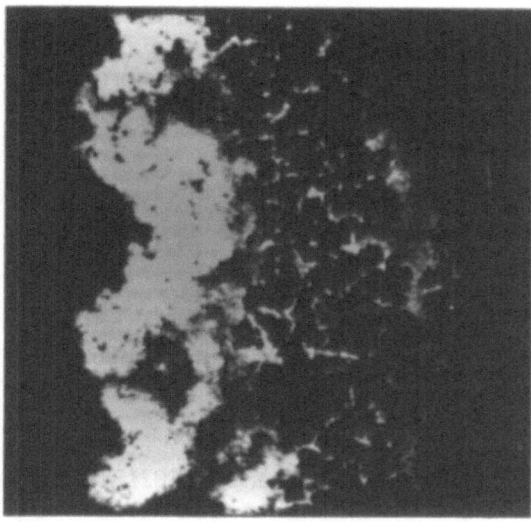

Fig. 2. In vitro generation of lipid peroxidation products in rat retinas. Eye cups from naive rats were incubated aerobically with the radical chain initiator 2,2′-azo-bis(2-amidinopropane) hydrochloride for 2h at 37°C; and the cryostat sections were reacted with 3-hydroxy-2-naphthoic acid hydrazide and p-toluenesulfonic acid as described for in vivo experiments (×460)

At physiologic pH, peroxynitrite has been shown to nitrate especially the phenolic rings, forming nitrotyrosine from tyrosine; this reaction forms the basis for the detection scheme used in the present study. The other common aromatic amino acid, phenylalanine, is poorly nitrated by the peroxynitrite due to the absence of activating hydroxyl groups in the aromatic ring. The abundance of nitrotyrosine staining in an area is presumably governed by the concentration of the peroxynitrite generated and by the availability of tyrosine molecules in that particular area. The quantity of tyrosine in the various regions of the retina, however, has not been reported in the past, possibly because of the difficulty quantitating a particular amino acid in a pool of protein.

The initiation of membrane lipid peroxidation by peroxynitrite has been demonstrated previously. For example, peroxynitrite was found to decompose to hydroxyl radical-like species and nitrogen dioxide, initiating lipid peroxidation without a need for iron (Radi et al. 1991). The hydroxyl radical-like species are apparently capable of extracting hydrogen atoms from polyunsaturated fatty acids (PUFAs) in a manner similar to that of true hydroxyl radicals (Radi et al. 1991). Unlike other organs, photoreceptor membranes contain abundant PUFAs; docosahexaenoic acid (22:6, omega 3), especially, accounts for nearly 50% of the total fatty acid content. These PUFAs are especially prone to peroxidation in the presence of suitable initiators. Once formed, fatty acid hydroperoxides display chemotactic activity toward PMNs, further perpetuating the inflammation (Goto et al. 1991).

In the past we demonstrated that there was neutrophil-mediated retinal damage in EAU. The extent of membrane lipid peroxidation was quantitated by extracting the oxidation product and measuring the parameters, including conjugated dienes, ketodienes, thiobarbituric acid-reactive substances, and fluorescent Schiff-base compounds (Goto et al. 1992; Wu et al. 1992). This study, therefore, not only confirms the occurrence of EAU-mediated retinal lipid peroxidation; it also determined that the primary site of production of these products is at the photoreceptors.

Conclusions

The present study reveals the presence of peroxynitrite in EAU. Moreover, the site of peroxynitrite production appears to coincide with the site of oxidative damage in the retina. Peroxynitrite-mediated damage to photoreceptors, therefore, may be one of the primary events leading to the subsequent pathological degeneration of photoreceptors in EAU.

Acknowledgment. Supported in part by grants EY12363 and EY10212 from the National Institutes of Health.

References

Allen JB, McGahan MC, Fleisher LN, Privalle CT (1996) Intravitreal lipopolysaccharide (LPS) induces peroxynitrite formation (ARVO abstract). Invest Ophthalmol Vis Sci 37:S918

Babbs CF, Griffin DW (1989) Scatchard analysis of methane sulfinic acid production from dimethyl sulfoxide: a method to quantity hydroxyl radical formation in physiologic systems. Free Radic Biol Med 6:493–503

Goto H, Wu GS, Chen F, Kristeva M, Sevanian A, Rao NA (1992) Lipid peroxidation in experimental uveitis: sequential studies. Curr Eye Res 11:489–499

Goto H, Wu GS, Gritz DC, Attala LR, Rao NA (1991) Chemotactic activity of the peroxidized retinal membrane lipids in experimental autoimmune uveitis. Curr Eye Res 10:1009–1014

Gritz DC, Montes C, Attala LR, Wu GS, Sevanian A, Rao NA (1991) Histochemical localization of superoxide production in experimental autoimmune uveitis. Curr Eye Res 10:927–931

Radi R, Beckman JS, Bush KM, Freeman BA (1991) Peroxynitrite-induced membrane lipid peroxidation: the cytotoxic potential of superoxide and nitric oxide. Arch Biochem Biophys 288:481–487

Rao NA (1990) Role of oxygen free radicals in retinal damage associated with experimental uveitis. Trans Am Ophthalmol Soc 88:797–850

Weiss SJ, LoBuglio AF (1982) Phagocyte-generated oxygen metabolites and cellular injury. Lab Invest 47:5–18

Wu GS, Sevanian A, Rao NA (1992) Detection of retinal lipid hydroperoxides in experimental uveitis. Free Radic Biol Med 12:19–27

Zhang J, Wu GS, Rao NA (1993) Role of nitric oxide in experimental autoimmune uveitis (ARVO abstract). Invest Ophthalmol Vis Sci 34:1000

V. The Retina

Regulation of Inducible Nitric Oxide Synthase in the Retina

Olivier Goureau[1], Violaine Faure[1], Yvonne De Kozak[2],
Francine Behar-Cohen[1], Pablo Dighiero[1], and Yves Courtois[1]

Introduction

Nitric oxide (NO) is known to be synthesized enzymatically in a tightly regulated manner by a number of tissues and cell types. The NO produced by cells serves a wide variety of functions in such cells as vascular endothelia, immune cells, neurons and glia, hepatocytes, and smooth muscle cells (for review, Bredt and Snyder 1994; Knowles and Moncada 1994; Nathan and Xie 1994). The functions of NO are diverse, having actions on vascular tone, neurotransmission (Bredt and Snyder 1994), immune cytotoxicity (Nüssler and Billiar 1993), and others.

Three isoforms of NO synthase (NOS) have been identified as being responsible for this synthesis in the presence of oxygen, NADPH, and flavins; they represent three distinct gene products (Knowles and Moncada 1994; Nathan and Xie 1994). Two of the enzyme types are continuously present and thus are termed constitutive NOS. The first, termed NOS-I, is found in the cytosol of central and peripheral neurons (Bredt and Snyder 1994); the second (NOS-III) was originally expressed by the vascular endothelium. Small amounts of NO are generated by these two isoenzymes when they are activated by the calcium–calmodulin complex. In contrast, a wide array of cell types can express the inducible form of NOS (NOS-II or iNOS) when stimulated by endotoxins or cytokines (Nathan and Xie 1994). This expression was first demonstrated in rodent macrophages and then extended in vitro to other cells (Nüssler and Billiar 1993; Nathan and Xie 1994). This isoform, the activity of which is independent of calcium and calmodulin, generates large amounts of NO over longer periods and is dependent on the presence of the stimuli.

Since the early 1990s the NO pathway has also been studied in the retina, and the inducible and constitutive isoforms have been identified (for review: Goldstein et al. 1996; Goureau and Courtois 1996).

[1]Développement, Vieillissement et Pathologie de la Rétine, INSERM U450, 29 rue Wilhem, 75016 Paris, France
[2]Laboratoire d'Immunopathologie Oculaire, INSERM U86, 15 rue de l'Ecole de Médecine, 75270 Paris 06, France

Expression of NOS-II in Retinal Cells In Vitro

We have demonstrated that retinal müller glial (RMG) cells can express NOS-II after endotoxin and cytokine stimulation (Goureau et al. 1994a). Bovine (Goureau et al. 1992), human (Goureau et al. 1994b), and murine (Liversidge et al. 1994; Sparrow et al. 1994) retinal pigmented epithelial (RPE) cells also contain NOS-II. In rat and bovine RPE cells NOS-II mRNA and enzyme activity are induced by synergistic cooperation between interferon γ (IFNγ) and lipopolysaccharide (LPS) and can be potentiated by tumor necrosis factor α (TNFα); costimulation with IFNγ and interleukin-1β is essential for NO production in human RPE cells. We have also demonstrated that fibroblast growth factors (FGFs) and transforming growth factor β (TGFβ) have opposing actions on the regulation of the production of NO by RPE cells (Goureau et al. 1993, 1994b). In bovine RPE cells, FGF-1 and FGF-2 inhibit the induction of NOS at the transcriptional level (Goureau et al. 1995a), but it is not an inhibitor of NOS-II induction in rat or human RPE cells or rat RMG cells (Goureau et al. 1994a, 1994b). The opposite data were observed with TGFβ, which frequently acts as an immunosuppressor signal, as it largely prevents NOS-II induction in murine RPE and RMG cells (Goureau et al. 1994a, Sparrow et al. 1994).

We demonstrated that treatment of RPE cells with IFNα or IFNβ markedly reduced the production of nitrite due to LPS and IFNγ (Table 1). The inhibitory effect of IFNα and IFNβ on NOS-II activity was correlated with a decrease of NOS-II protein and mRNA accumulation (data not shown). Because NO might act as an antiviral effector molecule in the retina, the production of IFNα and IFNβ during viral infection might weaken the antiviral defense in the retina by preventing NOS-II induction and a large release of NO.

We have attempted to elucidate the mechanism involved in the induction of NOS-II by investigating the transduction pathways leading to activation of the NOS-II gene by LPS and IFNγ in bovine RPE cells. It has been established that LPS and IFNγ activated tyrosine kinases and reactive oxygen species that activated specific transcription factors (Shuai 1994; Ulevitch and Tobias 1994). Among these factors, the nuclear transcription factor κB (NF-κB) and interferon regulatory factor 1 (IRF-1), respectively activated by LPS and IFNγ, were implicated in the induction of NOS-II (Martin et al. 1994; Xie et al. 1994).

In bovine RPE cells we observed that addition of a tyrosine kinase inhibitor, genistein, or an antioxidant such as pyrrolidinedithiocarbamate (PDTC), markedly decreased the nitrite release due to LPS and IFNγ (Table 1). These results demonstrated that the mechanism involved in the induction of NOS-II used tyrosine kinases and production of a reactive oxygen species. Furthermore, the activation of NF-κB by LPS could be blocked by addition of PDTC (data not shown), confirming the role of this transcription factor in NOS-II induction. Preliminary experiments reported that IFNγ can rapidly induce IRF-1 mRNA in RPE cells, suggesting that this factor could be also involved in the regulation of NOS-II expression in bovine RPE cells.

TABLE 1. Inhibition of LPS/IFNγ-induced nitrite release by IFNα, tyrosine kinase inhibitors, and antioxidants

Addition	Nitrite (μM)
None	<2
LPS (1 μg/ml) + IFNγ (100 U/ml)	21 ± 2.8
LPS + IFNγ + IFNα (100 U/ml)	3.1 ± 1.9
LPS + IFNγ + genistein (25 μg/ml)	5.7 ± 0.7
LPS + IFNγ + PDTC (10 μM)	<2

Confluent RPE cells were incubated for 72 hours in medium alone or with LPS and IFNγ with or without IFNα, PDTC, or genistein. Nitrite release was determined in culture supernatant by the Griess reaction.
LPS, lipopolysaccharide; IFN, interferon; PDTC, pyrrolidinedithiocarbamate; RPE, retinal pigmented epithelial.

In Vivo Expression of NOS-II in Retinal Pathology

We investigated the possibility that NO could be involved in the development of retinal pathologies, particularly some inflammatory diseases. In this context we demonstrated that NOS-II can be expressed in vivo in human retina as a result of viral infection (Dighiero et al. 1994). Immunohistochemistry studies revealed that NOS-II was detected in cytomegalovirus (CMV)-infected retina from patients with acquired immunodificiency syndrome (AIDS) whereas no staining was observed in the control patient (Fig. 1). We also demonstrated that NOS-II was localized in CMV-infected glial cells (Dighiero et al. 1994). The role of NO in viral infections of the retina could be beneficial, via its antimicrobial and antiviral effects, but also detrimental through its potential to damage tissue.

We have reported that subcutaneous injection of endotoxin in the rat dramatically induces release of nitrite in the aqueous humor and vitreous 16 h after the injection (Goureau et al. 1995b). This nitrite release could be partially prevented by an intraperitoneal injection of N^G-nitro-L-arginine methyl ester (L-NAME), an inhibitor of NOS (Goureau et al. 1995b). Reverse transcription polymerase chain reaction (RT-PCR) analysis demonstrated that NOS-II mRNA could be detected in the anterior segment of the eye and the retina 16 h after endotoxin injection (Fig. 2). The cells that are the primary source of NO in the retina have been determined by in situ hybridization and correspond to resident cell types (glial cells) and infiltrating cells (polymorphonuclear macrophages) (Jacquemin et al. 1996). These results support the hypothesis that both inflammatory and resident ocular cells are involved in NOS-II expression during endotoxin-induced uveitis (EIU). Using the EIU model, we reported that intraperitoneal injections of the NOS inhibitor L-NAME, which prevents ocular NO release, inhibited clinical inflammation in the anterior and posterior parts of the eye (Goureau et al. 1995b). This result confirmed the potential role of NO in EIU in Lewis rats, as suggested by others (Mandai et al. 1994; Parks et al. 1994).

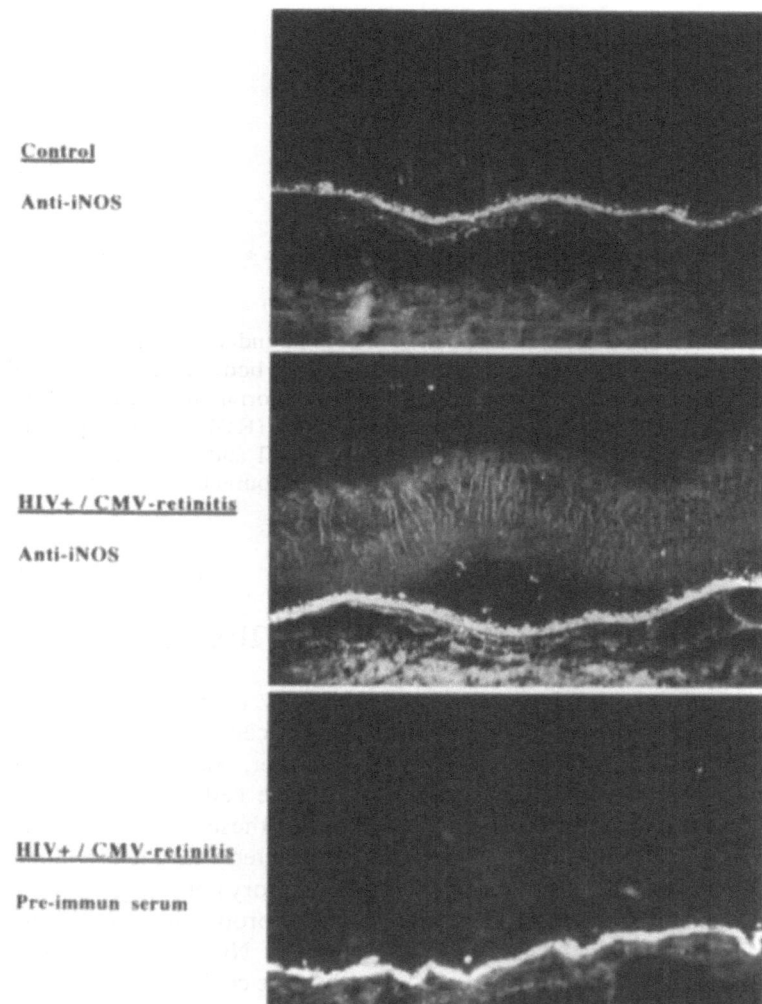

Control

Anti-iNOS

HIV+ / CMV-retinitis

Anti-iNOS

HIV+ / CMV-retinitis

Pre-immun serum

FIG. 1. Presence of nitric oxide synthase II (NOS-II) in cytomegalovirus (CMV)-infected retina from patients with acquired immunodificiency syndrome (AIDS). Specific immuno-histochemistry shows NOS-II (iNOS) in retinas from AIDS patients affected with CMV retinitis-specific immunolabeling, whereas no signal is detected in the retina from a seronegative patient. Control experiments with preimmunization serum in a retinitis-affected AIDS patient gave negative results. (Adapted from Dighiero et al. 1994)

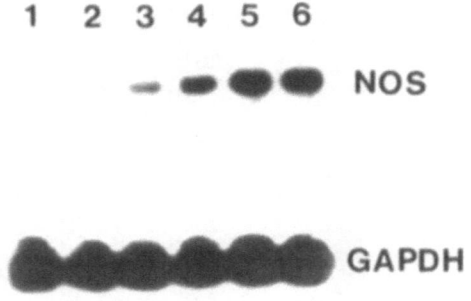

FIG. 2. Expression of NOS-II mRNA during endotoxin-induced uveitis. RNA extracted from the retina (*lanes 1, 3, 4*) and from the iris/ciliary body (*lanes 2, 5, 6*), 16 h after lipopolysaccharide (LPS) injection (*lanes 3–6*) or from normal rats (*lanes 1, 2*) was used for each reverse transcriptase-polymerase chain reaction (RT-PCR). PCR products were identified using specific hybridization probes for NOS-II and reduced glyceraldehyde-phosphate dehydrogenase (GAPDH). (Adapted from Goureau et al. 1995b)

Importance of Peroxynitrite in Retinal Toxicity

The NO alone seems unable to induce cell toxicity, as in LPS/IFNγ-stimulated RPE cells, which produce NO, we failed to detect cell damage (Goureau et al. 1993). Furthermore, addition of an NO donor, such as 3-morpholino-sydnonimine (SIN-1), to RPE cells did not induce cell death (Becquet et al. 1994). In vivo experiments are in agreement with these in vitro results, as any ocular damage could be detected in rabbit eye injected with the NO donor SIN-1 (Behar-Cohen et al. 1996a). With ocular inflammatory pathologies, such as EIU, a release of superoxide anion in parallel with NO production could lead to the formation of peroxynitrite via the combination of NO and superoxide anion (Beckman et al. 1990). Peroxynitrite is a highly toxic compound that can initiate lipid peroxidation, protein oxidation, and tyrosine nitration (Beckman and Crow 1993).

To determine the toxicity of peroxynitrite in the retina, we investigated in vitro the effect of peroxynitrite addition to bovine RPE cells. Cell viability experiments demonstrated that 6 h after peroxynitrite treatment only 40% of RPE cell survive (Table 2). This effect is due to the peroxynitrite, as no cell damage was observed with the decomposed peroxynitrite (Table 2). We have also demonstrated that addition of peroxynitrite induced RPE cell apoptosis, determined by Tunel assay (Behar-Cohen et al. 1996b), suggesting that peroxynitrite-induced cell death could be mediated at least in part by apoptosis. Furthermore, immuno-histochemistry studies with a nitrotyrosine antibody have demonstrated the presence of such protein modifications in RPE cells treated with peroxynitrite for 6 h but not in control RPE cells (Fig. 3). The significance of such nitration and

TABLE 2. Toxicity induced by peroxynitrite on RPE cells

Addition	Cell viability (% of control)
None	100
Peroxynitrite (0.5 mM)	52 ± 2
Peroxynitrite (2.5 mM)	18 ± 1.5
Decomposed peroxynitrite (2.5 mM)	95 ± 3

Cells were incubated with peroxynitrite at the indicated concentrations. Cell viability was evaluated after 24 h using crystal violet nuclear staining.

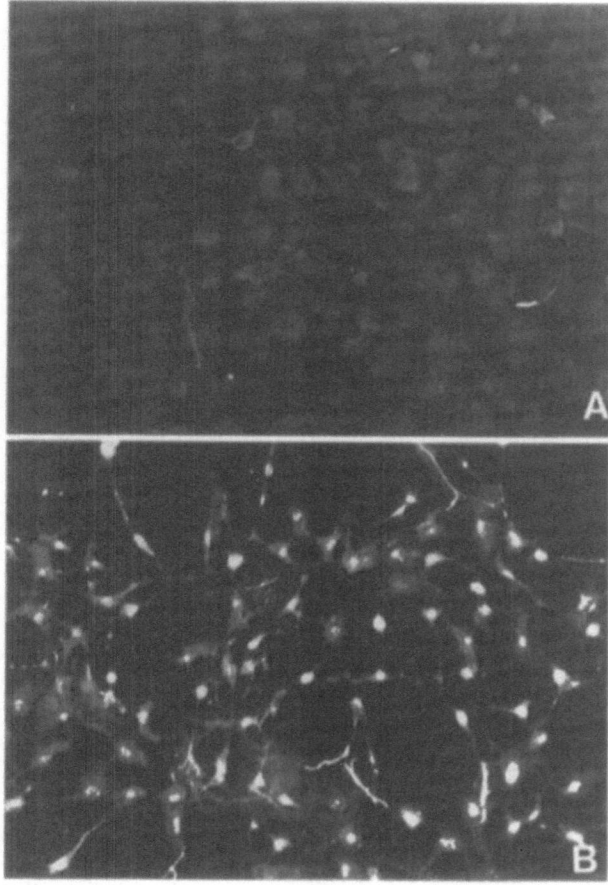

FIG. 3. Detection of nitrotyrosine after retinal pigmented epithelial (RPE) cells were treated with peroxynitrite. No staining with antinitrotyrosine was observed in control RPE cells (**A**), whereas intense fluorescence appeared in cells 6 h after peroxynitrite treatment (**B**). (Adapted from Behar-Cohen et al. 1996b)

its relation to the apoptic process and cell death are currently under investigation by attempting to identify the molecular targets that are nitrated in RPE cells. It would be tempting to postulate that peroxynitrite, more likely than NO itself, is implicated in the pathogenesis of various retinal damage in vivo.

References

Beckman JS, Crow JP (1993) Pathological implications of nitric oxide, superoxide and peroxynitrite formation. Biochem Soc Transact 21:330–334

Beckman JS, Beckman TW, Chen J, Marshall PA, Freeman BA (1990) Apparent hydroxyl radical production by peroxynitrite: implications for endothelial injury from nitric oxide and superoxide. Proc Natl Acad Sci USA 87:1620–1624

Becquet F, Courtois Y, Goureau O (1994) Nitric oxide decreases in vitro phagocytosis of photoreceptor outer segments by bovine retinal pigmented epithelial cells. J Cell Physiol 159:256–262

Behar-Cohen F, Goureau O, D'Hermies F, Courtois Y (1996a) Decreased intraocular pressure induced by nitric oxide donors is correlated to nitrite production in the rabbit eye. Invest Ophthalmol Vis Sci 37:1711–1715

Behar-Cohen F, Heydolph S, Faure V, Droy-Lefaix M, Courtois Y, Goureau O (1996b) Peroxynitrite cytotoxicity on bovine retinal pigmented epithelial cells in culture. Biochem Biophys Res Commun 226:842–859

Bredt DS, Snyder SH (1994) Nitric oxide: a physiologic messenger molecule. Annu Rev Biochem 63:175–195

Dighiero P, Reux I, Hauw JJ, Fillet AM, Courtois Y, Goureau O (1994) Expression of inducible nitric oxide synthase in cytomegalovirus-infected glial cells of retinas from AIDS patients. Neurosci Lett 166:31–34

Goldstein IM, Ostwald P, Roth S (1996) Nitric oxide: a review of its role in retinal function and disease. Vis Res 36:2979–2994

Goureau O, Courtois Y (1996) Nitric oxide in the retina: a double face mediator. Med Sci 12:593–598

Goureau O, Lepoivre M, Courtois Y (1992) Lipopolysaccharide and cytokines induce a macrophage-type of nitric oxide synthase in bovine retinal pigmented epithelial cells. Biochem Biophys Res Commun 186:854–859

Goureau O, Lepoivre M, Becquet F, Courtois Y (1993) Differential regulation of inducible nitric oxide synthase by basic fibroblast growth factors and transforming growth factor β in bovine retinal pigmented epithelial cells: inverse correlation with cellular proliferation Proc Natl Acad Sci USA 90:4276–4280

Goureau O, Hicks D, Courtois Y (1994a) Human retinal pigmented epithelial cells produce nitric oxide in response to cytokines. Biochem Biophys Res Commun 198:120–126

Goureau O, Hicks D, Courtois Y, de Kozak Y (1994b) Induction and regulation of nitric oxide synthase in retinal müller glial cells J Neurochem 63:310–317

Goureau O, Faure V, Courtois Y (1995a) Fibroblast growth factors decrease inducible nitric oxide synthase mRNA accumulation in bovine retinal pigmented epithelial cells. Eur J Biochem 230:1046–1052

Goureau O, Bellot J, Thillaye B, Courtois Y, de Kozak Y (1995b) Increased nitric oxide production in endotoxin-induced uveitis. J Immunol 154:6518–6523

Jacquemin E, de Kozak Y, Thillaye B, Courtois Y, Goureau O (1996) Expression of inducible nitric oxide synthase in the eye from endotoxin-induced uveitis in rats. Invest Ophthalmol Vis Sci 37:1187–1196

Knowles RG, Moncada S (1994) Nitric oxide synthases in mammals. Biochem J 298: 249–258

Liversidge J, Grabowski P, Ralston S, Benjamin N, Forrester JV (1994) Rat retinal pigment epithelial cells express an inducible form of nitric oxide synthase and produce NO in response to inflammatory cytokines and activated T cells. Immunology 83:404–409

Mandai M, Yoshimura N, Yoshida M, Iwaki M, Honda Y (1994) The role of nitric oxide synthase in endotoxin-induced uveitis: effect of N^G-nitro-L-arginine. Invest Ophthalmol Vis Sci 35:3673–3681

Martin E, Nathan C, Xie Q (1994) Role of interferon regulatory factor 1 in induction of nitric oxide synthase. J Exp Med 180:977–984

Nathan CF, Xie Q (1994) Nitric oxide synthases: roles, tolls, and controls. Cell 78:915–918

Nüssler AK, Billiar TR (1993) Inflammation, immunoregulation, and inducible nitric oxide synthase. J Leukoc Biol 54:171–178

Parks DJ, Cheung MK, Chan CC, Roberge FG (1994) The role of nitric oxide in uveitis. Arch Ophthalmol 112:544–546

Shuai K (1994) Interferon-activated signal transduction to the nucleus. Curr Opin Cell Biol 6:253–259

Sparrow JR, Nathan CF, Vodovotz Y (1994) Cytokine regulation of nitric oxide synthase in mouse retinal pigment epithelial cells in culture. Exp Eye Res 59:129–139

Ulevitch RJ, Tobias PS (1994) Recognition of endotoxin by cells leading to transmembrane signaling. Curr Opin Cell Biol 6:125–130

Xie Q, Kashiwabara Y, Nathan C (1994) Role of transcription factor NFκB/Rel induction of nitric oxide synthase. J Biol Chem 269:4705–4708

Neuronal Protection by Nitric Oxide-Related Species

Stuart A. Lipton, Yun-Beom Choi, Nikolaus J. Sucher, and H.S.-Vincent Chen

Redox Modulation and NO-Related Species

As endogenous sources of oxidizing and reducing agents have been discovered, redox modulation of protein function has been recognized to be an important mechanism for many cell types. For our purposes, we confine our review of redox modulation to covalent modification of sulfhydryl (thiol) groups on protein cysteine residues with special reference to the N-methyl-D-aspartate (NMDA) subtype of glutamate receptor in the brain. If the cysteine sulfhydryls possess a sufficient redox potential, oxidizing agents can react to form adducts on single sulfhydryl (thiol, -SH) groups; or if two free sulfhydryl groups are vicinal (in close proximity), disulfide bonds may be formed. Reducing agents can regenerate free sulfhydryl (thiol, -SH) groups by donating electron(s). Considering endogenous redox agents, in addition to the usual suspects including glutathione, ascorbate, vitamin E, lipoic acid, and reactive oxygen species, nitric oxide (NO) and its redox-related species have come to the fore. This has occurred largely because of the rediscovery and application to biological systems of work from the early part of the twentieth century showing the organic synthesis of nitrosothiols (RS-NO) (for review, Stamler et al. 1992). NO group donors represent different redox-related species of the NO group, each with its own distinctive chemistry, that lead to entirely different biological effects. NO-related species include nitric oxide (NO·) but also the other redox-related forms of the NO group: with one less electron (NO^+, or nitrosonium ion) or one additional electron (NO^-, or nitroxyl anion) (Stamler et al. 1992). Evidence suggests that all three of these redox-related forms or their functional equivalents are important pharmacologically and physiologically, participating in distinctive chemical reactions.

CNS Research Institute, Brigham and Women's Hospital, and Program in Neuroscience, Harvard Medical School, Boston, MA 02115, USA; Center for Neuroscience and Aging, The Burnham Institute, La Jolla, CA 92037, USA

Reaction of Cysteine Sulfhydryls with the NO Group

Free endogenous nitrosonium (NO^+) exists only at low pH. However, functional equivalents of NO^+ can be transferred to thiol or, more properly, thiolate anion (RS^-) at physiological pH. For example, transfer of NO^+ equivalents occur from one nitrosothiol to another, a reaction termed transnitrosylation (i.e., R-SH + R'-SNO $\rightleftarrows$ R-SNO + R'-SH). Because transfer of NO^+ equivalents involves thiolate anion ($R-S^-$) in contrast to thiol (R-SH), it is therefore pH-dependent (Arnelle and Stamler 1995). Endogenous nitrosothiols, such as S-nitroso-glutathione, have been demonstrated to exist in brain and lung and to react in this manner. The enzymatic machinery underlying the formation and breakdown of nitrosothiols is just beginning to be characterized. For example, thioredoxin reductase was shown to catalyze the homolytic cleavage of nitrosothiol (R-SNO) to nitric oxide ($NO\cdot$ + $RS\cdot$) (Nikitovic and Holmgren 1996).

The groups of Schmidt and Feelish presented evidence that neuronal nitric oxide synthase produces NO^- rather than $NO\cdot$ (Schmidt et al. 1996). NO^- presents an arcane chemistry, as NO^- can apparently be encountered in two states, consisting of either a high (singlet) or low (triplet) energy state, each with distinctive chemistries. In particular, singlet NO^- can react directly with thiol, whereas triplet NO^- does not (Bonner and Stedman 1996). In the triplet state NO^- may react with O_2 to form peroxynitrite ($ONOO^-$), which in turn may oxidize free thiols to disulfide (Radi et al. 1991; Kim et al. 1996).

Classically, there was no precedent for direct reaction of $NO\cdot$ with thiols (Pryor and Lightsey 1981; Pryor et al. 1982). However, Ischiropoulos and co-workers demonstrated that in the presence of an electron acceptor (e.g., O_2) $NO\cdot$ *can* react with thiol to form a nitrosothiol (Gow et al. 1997). One perhaps overly simplistic but useful conceptualization of this reaction is that the intermediate formed by an electron acceptor and $NO\cdot$ would be effecting NO^+ transfer in the presence of thiol. Nonetheless, the reaction of $NO\cdot$ and $O_2\cdot^-$ to form peroxynitrite would be kinetically favored over the formation of nitrosothiol if superoxide anion is present: for example, if $O_2\cdot^-$ is not scavenged by superoxide dismutase (SOD).

Another important concept when considering the possible chemical reactions of the NO group involves our image of the local diffusion and ephemeral nature of $NO\cdot$. Bredt and colleagues demonstrated that neuronal nitric oxide synthase (NOS) is located close to potential targets of NO by virtue of its PDZ domain (Brenman et al. 1996). For example, NOS interacts via its PDZ domain with the carboxyl-terminal tail of NMDAR1, the subunit of the N-methyl-D-aspartate (NMDA) receptor that is essential for functional activity. Therefore restricted diffusional constraints and the need for high local concentrations to facilitate NO reactions should not present a problem.

With some of the chemical reactions of these NO-related species in hand, we next pay particular attention to the mechanism of S-nitrosylation or transfer of the NO moiety to cysteine sulfhydryl group(s) on the NMDA receptor. It is of particular note that S-nitrosylation has also been shown to regulate the activity of various other ion channel proteins, G-proteins, growth factors, enzymes, and transcription factors. These reactions of NO-related species do not involve the well-known activation of guanylate cyclase by reaction with heme to increase cyclic guanosine monophosphate (cGMP) formation. Rather, they involve reactions with cysteine sulfhydryls on an increasing number of protein targets to provide modulation of function, analogous to the phosphorylation of critical serine, threonine, or tyrosine residues.

The chemical reactivity of the NO group and its associated redox states is related to the local redox milieu and peptide environment, pH, temperature, and the presence of catalytic amounts of transition metals. Here, we illustrate S-nitrosylation (transfer of NO^+ equivalents to thiol groups) to modulate protein functional activity. The first published example of this phenomenon is represented by the reaction of the NO group with regulatory sulfhydryl(s) of the NMDA receptor's redox modulatory site(s), resulting in down-regulation of receptor activity (Lipton et al. 1993). It had been known that NO donors could decrease NMDA function (Lei et al. 1992; Manzoni et al. 1992; Manzoni and Bockaert 1993), but the exact mechanism remained in question (Fagni et al. 1995). The redox basis for this reaction is presented below.

Most importantly, each of the NO-related species (NO^+, $NO\cdot$, and NO^- in singlet or triplet energy states) participates in different chemical reactions (Stamler et al. 1992; Lipton et al. 1993; Lipton and Stamler 1994). Nowhere is this more apparent than in the reactions of $NO\cdot$ versus NO^+, which influence neuronal survival in a diametrically opposed fashion (Lipton et al. 1993). Whereas $NO\cdot$ reacts with superoxide anion ($O_2\cdot^-$) to form peroxynitrite ($ONOO^-$), which in turn triggers neurotoxic reactions either by itself or via its breakdown products, NO^+ equivalents can react with a redox modulatory site(s) on the NMDA receptor to down-regulate the receptor's activity, which produces neuroprotection (Beckman et al. 1990; Dawson et al. 1991; Lipton et al. 1993).

S-Nitrosylation, NMDA Receptor Down-regulation, and Neuroprotection

The NO group can down-regulate NMDA receptor activity (Hoyt et al. 1992; Lei et al. 1992; Manzoni et al. 1992), apparently at a redox modulatory site(s) of the receptor, consisting of critical cysteine sulfhydryl or thiol group(s) (Lei et al. 1992; Lipton et al. 1993; Kohr et al. 1994; Sullivan et al. 1994). In native neurons (e.g., rat retinal ganglion cell neurons and cerebrocortical cells), we measured the amplitude of NMDA-evoked responses monitored by whole-cell recording with a patch electrode or by digital calcium imaging with the Ca^{2+}-sensitive dye fura-2 (Lei et al. 1992; Lipton et al. 1993). We found that sulfhydryl-reducing agents, such as dithiothreitol (DTT), which promote the formation of free thiol groups, increased NMDA responses. In contrast, oxidizing agents such 5,5′-dithiobisnitrobenzoic acid (DTNB) decreased NMDA responses by forming thiobenzoate protein conjugates at single sulfhydryl groups or perhaps by facilitating disulfide bond formation. Additionally, taken together with the DTT and DTNB results, we knew that thiols on the NMDA receptor were involved because under our conditions N-ethylmaleimide (NEM), a relatively specific agent for alkylating thiols, irreversibly blocked the effects of these redox reagents while itself slightly decreasing responses to NMDA (Lei et al. 1992; Lipton et al. 1993). Importantly, under specific conditions NEM also prevented the subsequent effects of NO donors, indicating that reactions of thiol and NO groups were involved. Our group and that of Bockaert (Manzoni and Bockaert 1993) have demonstrated that endogenous production of NO can decrease NMDA receptor activity, indicating the potential physiological importance of this effect. In these experiments implicating the involvement of endogenous NO, inhibition of NOS was found to enhance subsequent NMDA receptor responses.

As an example of NO chemical reaction at the NMDA receptor, we found that S-nitrosocysteine (cys-NO) decreases NMDA receptor activity as demonstrated by whole-cell recording or by digital calcium imaging (Lipton et al. 1993). In the presence of SOD, cys-NO attenuated NMDA-evoked Ca^{2+} influx, a prerequisite for NMDA receptor-mediated neurotoxicity. Not surprisingly therefore, under the same conditions application of cys-NO ameliorated NMDA receptor-mediated neurotoxicity. These findings can be explained best by cys-NO donating NO^+ equivalents. Thus S-nitrosylation or facile transfer of an NO^+ equivalent to thiol groups of the NMDA receptor (i.e., heterolytic fission of RS-NO) results in a nitrosothiol derivative of the NMDA receptor, which down-regulates receptor activity (Fig. 1). Under these conditions, any NO· produced by alternative homolytic cleavage of cys-NO is prevented from entering a neurotoxic pathway of peroxynitrite formation ($ONOO^-$) via reaction with $O_2^{.-}$ because of the presence of excess SOD (Lipton et al. 1993; Lipton and Stamler 1994). Rather, NO group transfer leads to down-regulation of NMDA receptor activity, possibly through facilitation of disulfide formation (Lipton et al. 1996a,b) (hence the dashed line in Fig. 1).

It is important to note that NO^+ transfer reactions may depend indirectly on catalytic amounts of transition metals. The fact that EDTA can prevent the effects of the NO group on NMDA receptor activity (Fagni et al. 1995) therefore supports rather than refutes this chemistry. In particular, metals can facilitate nitrosation reactions involving NO· (Stamler et al. 1992; Lipton et al. 1996a,b). Nitrosation of redox sites is facilitated by oxygen, transition metals, and perhaps $O_2^{.-}$ (superoxide anion) (Stamler 1994; Gow et al. 1997). The common event is transfer of an NO^+ equivalent or another intermediate with NO^+-like character to form an R-SNO.

In accordance with the report that nNOS may produce NO^- (nitroxyl anion) (Schmidt et al. 1996), we tested the ability of exogenous and endogenous generation of NO^- to modulate NMDA receptor function. We have found that donors of singlet NO^- can decrease NMDA receptor activity apparently via S-nitrosylation because the effects can be prevented by pretreatment with NEM. Additionally, enzymatic generation of NO^-, presumably in the triplet state,

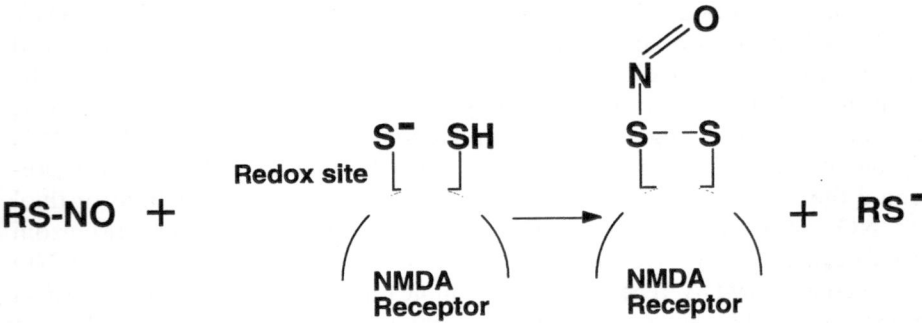

FIG. 1. *S*-Nitrosylation of critical thiol group(s) of the *N*-methyl-D-aspartate (NMDA) receptor's redox modulatory site by *S*-nitrosocysteine (cys-NO), a more general example of which is RS-NO, a nitrosylated protein. The redox modulatory site of the NMDA receptor is transnitrosylated by transferring the NO group (in the NO⁺ form) from RS-NO to cysteine sulfhydryl group(s) on the NMDA receptor-channel complex. It results in a decreased frequency of channel opening and hence decreased NMDA receptor activity

can also decrease NMDA responses but apparently via formation of peroxynitrite following reaction of triplet NO^- with O_2 (Kim et al. 1996; Lipton et al. 1996a,b).

Our work on nitrosylation and other redox reactions of recombinant NMDA receptors in the *Xenopus* oocyte expression system is instructive but must be interpreted with a degree of caution (Sullivan et al. 1994; Sucher et al. 1996). We do not yet appreciate how to form recombinant NMDA receptors that exactly mimic native receptors, and therefore conclusions based on site-directed mutagenesis studies of cysteines must be viewed with tempered enthusiasm. In fact, during the course of performing PCR reactions based on primers containing the cysteines known to be unique to NMDA receptor subunits, our group discovered a new NMDA receptor subunit (now termed NR3A) (Ciabarra et al. 1995; Sucher et al. 1995). Additional unidentified NMDA subunits probably remain to be identified. Thus it is not yet possible to understand definitively the native NMDA receptor responses based on recombinant subunits. This statement notwithstanding, our preliminary data suggest that the cysteines at position 744 and 798 are important not only to redox reactions in general but also to the effect of NO on the NMDA receptor; however, additional cysteines on the NR2A subunit also contribute to the NO effect. This work is still in progress, but our preliminary results show that (1) specific NMDA receptor subunit combinations manifest larger NO-induced decreases in activity than other receptor subunit combinations (Omerovic et al. 1995; Sucher et al. 1996), and (2) at five cysteines on NR1 and NR2A underlie the NO effect on the NMDA receptor.

Nitroglycerin Down-regulation of NMDA Receptor Activity and Amelioration of Neurotoxicity In Vitro and In Vivo

Based on the above findings, the ideal NO group donor drug would be one that reacts readily with the critical thiol group(s) of the redox modulatory site(s) of the NMDA receptor to inhibit excessive Ca^{2+} influx. We therefore studied nitroglycerin (NTG) as an exemplary compound. Specifically, this drug does not spontaneously liberate true nitric oxide (NO·) to any significant extent, and it is known to react readily with thiol groups forming derivative thionitrites (RS-NO) or thionitrates (RS-NO$_2$) (together these substances are represented as RS-NO$_x$) (Lei et al. 1992; Lipton et al. 1993).

Using whole-cell recording via patch-clamp electrodes and digital calcium imaging with fura-2, we found that nitroglycerin inhibited NMDA-evoked currents and Ca^{2+} influx (Lei et al. 1992; Lipton et al. 1993). Strong evidence that this effect of nitroglycerin is mediated by its reactions with thiol in the above-illustrated manner came from a series of chemical experiments, which showed that specific alkylation of thiol groups with NEM completely abrogated the inhibitory effect of nitroglycerin on subsequent NMDA-evoked responses under our conditions (Lei et al. 1992).

The finding that nitroglycerin could inhibit NMDA-evoked responses was corroborated by the demonstration that nitroglycerin also significantly ameliorates NMDA-induced neuronal killing in cerebrocortical and retinal ganglion cell cultures (Lei et al. 1992; Lipton et al. 1993). In addition, preliminary data suggest that high doses of nitroglycerin are neuroprotective in rat models of focal ischemia under conditions of constant systemic blood pressure and modestly increased cerebral blood flow in the penumbra (Lipton and Yang 1996). These parameters are held stable by either inducing tolerance to the systemic effects of nitroglycerin through chronic transdermal application (Sathi et al. 1993) or intravenous infusion of a pressor agent concurrently with nitroglycerin (Lipton and Yang 1996). Although difficult to prove in vivo, it appears likely that the decrease in stroke size observed after treatment with nitroglycerin is at least in part due to its effect on decreasing NMDA receptor activity, although other beneficial actions are also possible (Lipton and Yang 1996).

S-Nitrosylation of Cysteine Sulfhydryls on Other Ion Channels, Enzymes, Transcriptions Factors, and Regulatory Proteins

Shortly after NMDA receptor activity was shown to be regulated by NO-related species, similar data were presented for the Ca^{2+}-activated K^+ channel of cardiac muscle (Bolotina et al. 1994). In this case, donors of NO^+ equivalents were shown to activate the channel; and similar to findings at the NMDA receptor in our laboratory, NEM blocked the effect by irreversibly alkylating thiol groups.

Along these lines, several other ion channels, enzymes, G-proteins, transcription factors, and other proteins are either up-regulated or down-regulated by similar mechanisms of S-nitrosylation or donation of NO^+ equivalents to regulatory sulfhydryl centers (Stamler et al. 1997). The list will undoubtedly grow just as in recent years phosphorylation, myristolation, and palmitoylation have become recognized as important biochemical processes for regulatory function. Interestingly, palmitoylation may be aimed at similar critical thiol group targets, resulting in thioester bond formation. In fact, on some proteins such as SNAP-25 it is possible that S-nitrosylation and palmitoylation may compete for the same sulfhydryl, possibly with different physiological outcomes (Hess et al. 1993).

In contrast to phosphorylation, however, in the case of S-nitrosylation evidence is accumulating that the critical cysteine residues may be located extracellularly, intracellularly, or possibly even within the putative membrane spanning region of a protein. From this point of view S-nitrosylation may offer additional versatility in the modes of control that can be exerted compared to phosphorylation and other better known posttranslational forms of modification.

Conclusion

The possible chemical reactions of the NO group are dictated by its redox state. In the case of NO^+ equivalents, this mechanism appears to involve S-nitrosylation and possibly further oxidation of critical thiols to disulfide bonds in the NMDA receptor's redox modulatory site(s) to down-regulate channel activity. Other data suggest that NO^-, probably in the singlet state, can react with critical sulfhydryl group(s) of the NMDA receptor to down-regulate its activity; in the triplet state NO^- may oxidize these NMDA receptor sulfhydryl groups by formation of an intermediate such as peroxynitrite (Lipton et al. 1996a,b).

It is becoming increasingly evident that, in addition to NMDA receptors, biological activities of many other proteins containing critical cysteine residues can be regulated by S-nitrosylation and other redox reactions, in a sense similar to the type of control exerted by phosphorylation (Lipton et al. 1993). A putative consensus nitrosylation motif has been identified by our group for a range of proteins known to be nitrosylated, and this motif exists in the NMDA receptor as well (Stamler et al. 1997). This type of chemical reaction may represent a new, ubiquitous pathway for the molecular control of protein function by potentially reactive sulfhydryl centers.

Acknowledgments. This review is based on work performed in a close collaboration between the laboratories of Stuart Lipton at Harvard Medical School and Jonathan Stamler at Duke Medical Center. We are grateful to the many members of the two laboratories who contributed to the work. This was a case in which the mutual ignorance of the two groups exactly complemented one another, resulting in true collaboration. The work was presented nearly simultaneously at three meetings—in Kyoto, Berlin, and Barcelona—and therefore three related versions of this manuscript were offered for the proceedings of these meetings.

References

Arnelle DR, Stamler JS (1995) NO$^+$, NO·, and NO$^-$ donation by S-nitrosothiols: implications for regulation of physiological functions by S-nitrosylation and acceleration of disulfide formation. Arch Biochem Biophys 318:279–285

Beckman JS, Beckman TW, Chen J, Marshall PA, Freeman BA (1990) Apparent hydroxyl radical production by peroxynitrite: implications for endothelial injury from nitric oxide and superoxide. Proc Natl Acad Sci USA 87:1620–1624

Bolotina VM, Najibi S, Palacino JJ, Pagaon PJ, Cohen RA (1994) Nitric oxide directly activates calcium-dependent potassium channels in vascular smooth muscle. Nature 368:850–853

Bonner FT, Stedman G (1996) The chemistry of nitric oxide and redox-related species. In: Feelisch M, Stamler JS (eds) Methods in nitric oxide research. Wiley, Chichester, pp 3–18

Brenman JE, Chao DS, Gee SH, McGee AW, Craven SE, Santilliano DR, Wu Z, Huang F, Xia H, Peters MF, Froehner SC, Bredt DS (1996) Interaction of nitric oxide synthase with the postdynaptic density protein PSD-95 and α1-syntrophin mediated by PDZ domains. Cell 84:757–767

Ciabarra AM, Sullivan JM, Gahn LG, Pecht G, Heinemann S, Sevarino KA (1995) Cloning and characterization of χ-1: a developmentally regulated member of a novel class of the ionotropic glutamate receptor family. J Neurosci 15:6498–6508

Dawson VL, Dawson TM, London ED, Bredt DS, Snyder SH (1991) Nitric oxide mediates glutamate neurotoxicity in primary cortical cultures. Proc Natl Acad Sci USA 88:6368–6371

Fagni L, Olivier M, Lafon-Cazal M, Bockaert J (1995) Involvement of divalent ions in the nitric oxide-induced blockade of N-methyl-D-aspartate receptors in cerebellar granule cells. Mol Pharmacol 47:1239–1247

Gow AJ, Buerk DG, Ischiropoulos H (1997) A novel reaction mechanism for the formation of S-nitrosothiol in $vivo$. J Biol Chem 272:2841–2845

Hess DT, Patterson SI, Smith DS, Skene JHP (1993) Neuronal growth cone collapse and inhibition of protein fatty acylation by nitric oxide. Nature 366:562–565

Hoyt KR, Tang L-H, Aizenman E, Reynolds IJ (1992) Nitric oxide modulates NMDA-induced increases in intracellular Ca^{2+} in cultured rat forebrain neurons. Brain Res 592:310–316

Kim W-K, Rayudu PV, Mullins ME, Stamler JS, Lipton SA (1996) Down regulation of NMDA receptor activity in cortical neurons by peroxynitrite. In: Moncada S, Stamler JS, Gross S, Higgs EA (eds) The biology of nitric oxide. Part 5. Portland, London

Kohr G, Eckardt S, Lüddens H, Monyer H, Seeburg PH (1994) NMDA receptor channels: subunit-specific potentiation by reducing agents. Neuron 12:1031–1040

Lei SZ, Pan Z-H, Aggarwal SK, Chen H-SV, Hartman J, Sucher NJ, Lipton SA (1992) Effect of nitric oxide production on the redox modulatory site of the NMDA receptor-channel complex. Neuron 8:1087–1099

Lipton SA, Stamler JS (1994) Actions of redox-related congeners of nitric oxide at the NMDA receptor. Neuropharmacology 33:1229–1233

Lipton SA, Yang YF (1996) NO-related species can protect from focal cerebral ischemia/reperfusion. In: Krieglstein J (ed) Pharmacology of cerebral ischemia. Medpharm Scientific, Stuttgart, pp 183–191

Lipton SA, Choi Y-B, Pan Z-H, Lei SZ, Chen H-SV, Sucher NJ, Loscalzo J, Singel DJ, Stamler JS (1993) A redox-based mechanism for the neuroprotective and neurodestructive effects of nitric oxide and related nitroso-compounds. Nature 364:626–632

Lipton SA, Choi Y-B, Sucher NJ, Pan Z-H, Stamler JS (1996a) Redox state, NMDA receptors, and NO-related species. Trends Pharmacol Sci 17:186–187

Lipton SA, Kim W-K, Rayudu PV, Asaad W, Arnelle DR, Stamler JS (1996b) Singlet and triplet nitroxyl anion (NO⁻) lead to N-methyl-D-aspartate (NMDA) receptor down-regulation and neuroprotection. In: Stamler Gross JS, Moncada S (eds) The biology of nitric oxide. Portland, London

Manzoni O, Bockaert J (1993) Nitric oxide synthase activity endogenously modulates NMDA receptors. J Neurochem 61:368–370

Manzoni O, Prezeau L, Marin P, Deshager S, Bockaert J, Fagni L (1992) Nitric oxide-induced blockade of NMDA receptors. Neuron 8:653–662

Nikitovic D, Holmgren A (1996) S-Nitrosoglutathione is cleaved by the thioredoxin system with liberation of glutathione and redox regulating nitric oxide. J Biol Chem 271:19180–19185

Omerovic A, Chen S-J, Leonard JP, Kelso SR (1995) Subunit-specific redox modulation of NMDA receptros expressed in Xenopus oocytes. J Recept Signal Transduct Res 15:811–827

Pryor WA, Lightsey JW (1981) Mechanisms of nitrogen dioxide reactions: initiation of lipid peroxidation and the production of nitrous acid. Science 214:435–437

Pryor WA, Church DF, Govinden CK, Crank G (1982) Oxidation of thiols by nitric oxide and nitrogen dioxide: synthetic utility and toxicological implications. J Org Chem 47:156–159

Radi R, Beckman JS, Bush KM, Freeman BA (1991) Peroxynitrite oxidation of sulfhydryls: the cytotoxic potential of superoxide and nitric oxide. J Biol Chem 266:4244–4250

Sathi S, Edgecomb P, Warach S, Manchester K, Donaghey T, Stieg PE, Jensen FE, Lipton SA (1993) Chronic transdermal nitroglycerin (NTG) is neuroprotective in experimental rodent stroke models. Soc Neurosci Abstr 19:849

Schmidt HHHW, Holman H, Schindler U, Shutenko ZS, Cunningham DD, Feelisch M (1996) No ·NO from NO synthase. Proc Natl Acad Sci USA 93:14492–14497

Stamler JS (1994) Redox signaling: nitrosylation and related target interactions of nitric oxide. Cell 78:931–936

Stamler JS, Singel DJ, Loscalzo J (1992) Biochemistry of nitric oxide and its redox activated forms. Science 258:1898–1902

Stamler JS, Toone EJ, Lipton SA, Sucher NJ (1997) (S)NO signals: translocation, regulation, and a consensus motif. Neuron 18:691–696

Sucher NJ, Schahram A, Chi CL, Leclerc CL, Awobuluyi M, Deitcher DL, Wu MK, Yuan JP, Jones EG, Lipton SA (1995) Developmental and regional expression pattern of a novel NMDA receptor-like subunit (NMDAR-L) in the rodent brain. J Neurosci 15:6509–6520

Sucher NJ, Awobuluyi M, Choi Y-B, Lipton SA (1996) NMDA receptors: from genes to channels. Trends Pharmacol Sci 17:348–355

Sullivan JM, Traynelis SF, Chen H-SV, Escobar W, Heinemann SF, Lipton SA (1994) Identification of two cysteine residues that are required for redox modulation of the NMDA subtype of glutamate receptor. Neuron 13:929–936

Nitric Oxide and Retinal Ischemia

Satoshi Kashii[1], Yoshihito Honda[1], and Akinori Akaike[2]

Overview

If the entire retina of an experimental animal is subjected to ischemia for a certain period of time and is then reperfused, subsequent cell death occurs particularly in the inner part of the retina, manifesting in a delayed fashion (Fig. 1). This delayed neuronal death due to retinal ischemia is now believed to reflect impaired cellular ion homeostasis especially concerning Ca^{2+}.

Retinal ischemia induces a large increase in the release of glutamate, which exerts its toxic action by way of N-methyl-D-aspartate (NMDA) receptors on amacrine cells and retinal ganglion cells. Glutamate released during ischemia and especially during reperfusion first stimulates non-NMDA receptors and depolarizes the retinal neurons. As the membrane potentials are depolarized more from the resting membrane potentials, the blockade of NMDA receptors induced by Mg^{2+} is released. Ca^{2+} influx through the NMDA receptors activates a neuronal type of nitric oxide synthase (nNOS) situated in some amacrine cells, which produces nitric oxide (NO). There are three isoforms of NOS: nNOS, inducible NOS (iNOS) synthesized by macrophages, and endothelial NOS (eNOS). nNOS and eNOS produce NO only when the intracellular Ca^{2+} concentration is elevated and calmodulin is bound to the enzyme. By contrast, iNOS synthesizes NO continuously irrespective of the intracellular Ca^{2+}.

Once NO is generated, it diffuses out of the amacrine cells immediately. NO at low concentrations inhibits the NMDA receptors of the neighboring amacrine and ganglion cells as well as those producing NO; it thereby prevents further Ca^{2+} influx, which would lead to subsequent cell death. Hence, NO generated in

[1]Department of Ophthalmology and Visual Sciences, Graduate School of Medicine, Kyoto University, 54 Kawahara-cho, Shogoin, Sakyo-ku, Kyoto 606-8507, Japan
[2]Department of Pharmacology, Graduate School of Pharmaceutical Sciences, Kyoto University, Kyoto 606-8501, Japan

A   B   C   D

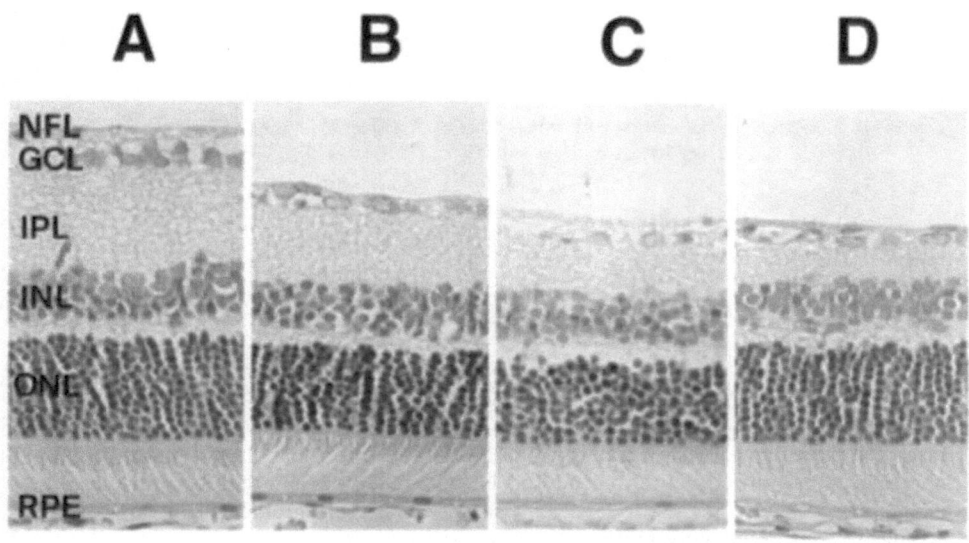

NFL
GCL

IPL

INL

ONL

RPE

50 μm

Fig. 1. Normal histology of the rat retina and delayed neuronal death after retinal ischemia. The histological section of an untreated rat retina demonstrates normal histology (**A**). Two vascular systems suppy blood to the retina. The inner two-thirds of the retina receives circulation via the central retinal artery, whereas the rest of the outer part of the retina is perfused with the choroid, which originates from the posterior ciliary arteries. The histological sections obtained from the rat retinas at 4 (**B**), 7 (**C**), and 14 (**D**) days after 60 min of ischemia induced by raising the intraocular pressure. The number of ganglion cells and thickness of the inner part of the retina decreased markedly with time after transient ischemia. *NFL*, nerve fiber layer; *GCL*, retinal ganglion cell layer; *IPL*, inner plexiform layer; *INL*, inner nuclear layer; *ONL*, outer nuclear layer; *RPE*, retinal pigment epithelium. Bar = 50 μm

small and highly regulated bursts by nNOS acts as a messenger molecule. By contrast, in pathological states of retinal ischemia, in which intracellular Ca^{2+} concentration remains persistently elevated owing to increased release of glutamate, nNOS is continuously active and produces high concentrations of NO; the NO interacts with oxygen radicals, becomes toxic, and mediates glutamate-induced retinal neuronal death (Fig. 2).

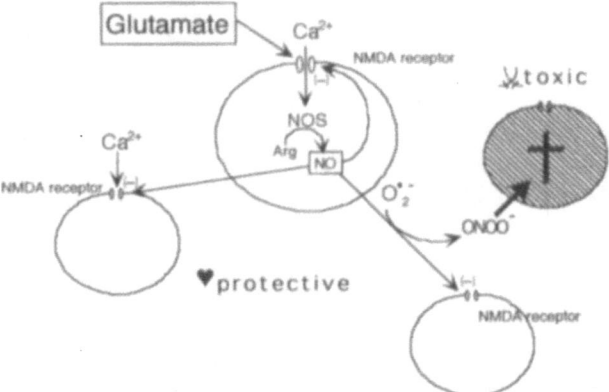

FIG. 2. Dual actions of nitric oxide (NO) in NMDA receptor-mediated neurotoxicity in the retina. NO at low concentrations inhibits NMDA receptors and thereby prevents glutamate neurotoxicity. NO is produced in excess, which then interacts with superoxide, becomes toxic, and mediates glutamate neurotoxicity

Blood Circulation of the Retina

Two vascular systems supply blood to the retina. The retinal vessels originating from the central retinal artery supply the inner two-thirds of the retina, extending from the inner aspect of the inner nuclear layer (INL) to the nerve fiber layer (NFL) over the retinal ganglion cell layer (GCL) (Fig. 1). The rest of the outer layers of the retina, extending from the outer aspect of the INL to the retinal pigment epithelium (RPE) including the outer nuclear layer (ONL) where photoreceptor cells are situated, are nourished from the choroid, which is derived from the posterior ciliary arteries. In humans and experimental animals such as rats and cats, the retina depends on both the retinal vessels and the choroid, whereas in some other animals (e.g., rabbits and guinea pigs) the choroidal capillaries are the primary source of the blood supply to the retina.

Clinical Aspects of Retinal Ischemia

"Retinal stroke" is a term used to describe a variety of retinal arterial occlusive diseases. Central retinal artery occlusion (CRAO), which usually occurs at the level of the lamina cribrosa just before the artery enters the retina (Hayreh 1971), is not an appropriate example of total retinal ischemia. A cherry-red spot, which is a pathognomonic sign of CRAO, develops because the fovea obtains its blood supply from the unaffected choroidal circulation. Electroretinography of CRAO reveals preservation of the a-wave and loss of the b-wave, substantiating the preserved function of the outer retina, which is supplied by the choroidal circulation. Total retinal ischemia develops only when obstruction occurs at the level of the ophthalmic artery and both the ciliary (i.e., choroidal) and retinal circulations are compromised at the same time. There are reports on such a case in the literature (Brown et al. 1986), but ophthalmic artery occlusion is rarely encountered in daily clinical practice.

Hedges in 1962 described a retinopathy associated with underlying, severe carotid obstructive desease characterized by insiduous onset, dilated and tortuous retinal veins, peripheral microaneurysms, and blossom-shaped hemorrhages in the midperipheral retina. Subsequently Kearns and Hollenhorst (1963) called the disorder "venous stasis retinopathy" and defined it as the ocular signs and symptoms occurring secondary to severe carotid artery obstruction. The same nomenclature has been used for a nonischemic form of central retinal vein occlusion described in 1971 by Hayreh that is an entity different in pathogenesis and clinical manifestation from that described by Kearns and Hollenhorst (Kearns 1983). Some patients with venous stasis retinopathy also manifest signs of reduced blood supply to the anterior segment. The combination of posterior and anterior ocular ischemia constitute ocular ischemic syndrome, which reflects inadequate ophthalmic artery perfusion. Based on the angiographic evidence of internal carotid artery occlusion combined with retrograde collateral flow to the brain through the ipsilateral external carotid and ophthlamic arteries, Miller (1991) suggested that the panocular ischemia that occurs in patients with carotid occlusive disease may not be the result of internal carotid artery occlusion per se but may be the result of an ocular steal mechanism in which the collateral circulation shunts blood away from the eye to the brain. Early retinopathy may resolve spontaneously with the development of collateral circulation in the retina. In eyes with persistent hypoperfusion, however, neovascularization develops throughout the ocular tissue including the iris, anterior chamber angle, retina, and optic disc; this situation predominates in disease states. The later development of neovascularization characterizes and distinguishes the ocular ischemia derived from brain ischemia. Although the exact incidence of the ocular ischemic syndrome is not known, Brown and his associates (1991) have estimated that approximately 1800 new cases (7.5 per million) would be encountered per year in the United States. Conversely, it has been estimated that 5%–20% of patients with carotid obstructive disease develop the ischemic ocular syndrome, including venous stasis retinopathy (Ross Russell and Page 1983; Brown et al. 1991).

Color Doppler analysis of ocular blood flow has demonstrated that blood flow in the central retinal artery and the posterior ciliary arteries is reduced early in the disease process of glaucoma before visual field defects can be detected (Nicolela et al. 1996). It has been reported that patients with primary open-angle glaucoma and normal-tension glaucoma have lower blood flow velocities and higher resistive indices in the central retinal artery, as well as shorter posterior ciliary arteries, than do normal control subjects (Rojanapongpun et al. 1993; Harris et al. 1994; Rankin et al. 1995). Nicolela et al. (1996) reported that eyes with normal visual fields and normal-appearing optic nerve disks in patients with asymmetrical glaucoma had reduced blood flow velocity and higher resistance in the central retinal artery and the short posterior ciliary arteries. This implies that the whole layer of the retina is subject to ischemia early in the course of the glaucomatous disease process. In contrast to ischemic ocular syndrome, however, glaucoma by no means causes neovascularization in the eye including the retina. The glaucomatous process may thus reflect a more degenerative aspect of the disease process induced by ischemic insults.

Experimental Retinal Ischemia

Experimentally, various methods have been used to study retinal ischemia, such as the panretinal ischemia induced by vascular ligation or the focal ischemia produced by photothrombosis. In the following experiments, we used a pressure-induction model to induce panretinal ischemia. Under sodium pentobarbital anesthesia (50 mg/kg i.p.), a 27-gauge needle is inserted into the anterior chamber of the experimental animal and connected to a bottle of irrigating solutions. Retinal ischemia is induced by elevating the intraocular pressure (IOP), which is regulated with the height of the bottle. According to our previous measurements of ocular blood flow in a male Sprague-Dawley rat (200 g) using a noncontact laser blood flowmeter, an IOP of 130 mmHg is optimal for inducing of panretinal ischemia and immediate recovery to baseline by releasing the raised IOP (Fig. 3). At each experiment, funduscopic examinations were always

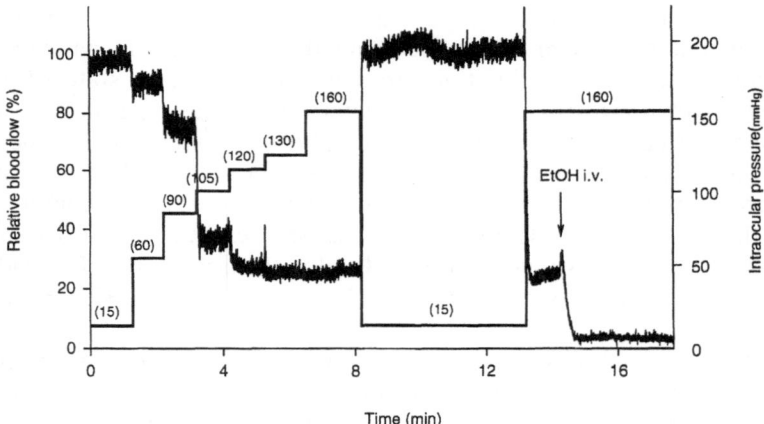

Fig. 3. Percentage changes in blood flow at various intraocular pressures (IOP) in a normal rat. Complete retinal ischemia was obtained at IOP 130 mmHg. Immediately after cession of ischemia by releasing the raised IOP down to 15 mmHg, the blood flow immediately returned to the preischemic baseline level. The numbers in parentheses indicate the IOP. Ethanol was administered intravenously at the end. (From Adachi et al. 1998b, with permission)

undertaken to ascertain visually the pallid chorioretinal structure and the attenuation and total blanching of retinal arterioles in each animal. Reperfusion of
the retinal vasculature was also evaluated funduscopically immediately after
reducing the IOP to 15mmHg. All animals were treated in accordance with the
Association for Research in Vision Science and Ophthalmology Resolution on
the Use of Animals in Research.

When histologically quantifying ischemic damage in the retina, one should
keep in mind that experimental conditions significantly affect the outcome following ischemia. Among these conditions, the body temperature of the experimental animal during retinal ischemia plays a critical role in the subsequent
histological change in the retina. We have previously demonstrated that the
reduction of the number of ganglion cells and thickness of the inner plexiform
layer (IPL) depends on the body temperature during ischemia (Fig. 4). Lowering the body temperature alone can offer significant protection to the retinal
ganglion cells and neurons in the INL (Adachi et al. 1998a). It was also reported
that local hypothermia applied to the rat eye via an icepack completely protected
the neurons in the INL from 120 min of ischemic injury caused by ligation of the
optic nerve. By contrast, eyes similarly subjected to 120 min of ischemia and

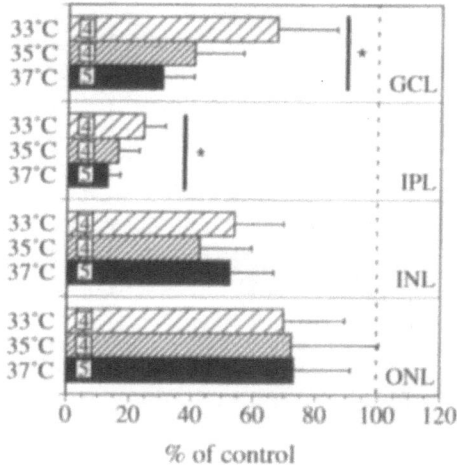

FIG. 4. Effect of the body temperature of the experimental animal on histological changes
after ischemia. Morphometric analysis was carried out on the histological sections
obtained 7 days after 60 min of ischemia. The rectal temperatures maintained during
surgery, including an ischemic period and recovery from anesthesia, are shown for each
bar. The following four parameters for the right eye, which was subjected to ischemia,
were normalized to those for the left, intact eye and are demonstrated as percent of
control: linear cell density (cells/mm) in the ganglion cell layer (*GCL*); thickness of the
inner plexiform layer (*IPL*), the inner nuclear layer (*INL*), and the outer nuclear layer
(*ONL*). The reduction of the cell density of the GCL and the thickness of the IPL were
dependent on body temperature. (From Adachi et al. 1998a, with permission)

48 h of reperfusion but without an ice pack showed severe damage (Faberowski et al. 1989). Therefore, strict maintenance of body temperature of an experimental animal during ischemica is mandatory. In the following discussion the data were obtained from rats whose rectal temperature was kept at 37°C using an electrical heating pad with a temperature probe and a thermostat until the animals recovered from anesthesia.

The selective vulnerability of inner retinal layers in ischemic injury has been investigated extensively (Adachi et al. 1998a). Here again, one must be aware of the duration of ischemia and reperfusion; that is, the acquisition time for obtaining histological sections significantly affects subsequent histological changes. Under our experimental settings, when histological sections were evaluated at day 7 after ischemia, there was time-dependent damage to the density of the GCL and the thickness of the IPL and INL following an ischemic period longer than 45 min (Fig. 5). However, the damage to the ONL became significant in terms of the thickness of that layer only if the duration of ischemia exceeded 60 min. Therefore, we elected to use 45 min as the duration of ischemia and

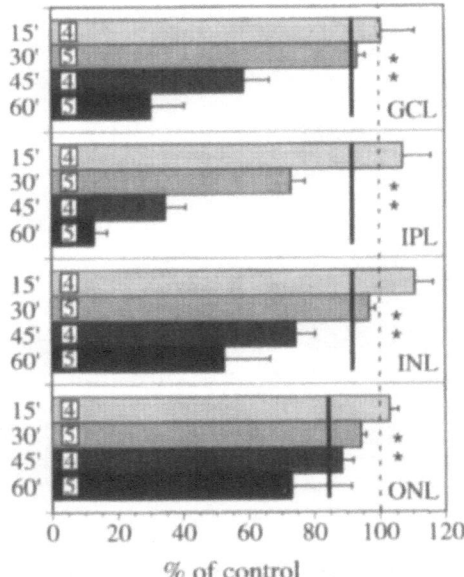

Fig. 5. Effect of the duration of ischemia on histological changes after ischemia. The body temperature of the experimental animal was maintained at 37°C. Histological sections were obtained 7 days after ischemia. The right eye was subjected to ischemia for 15–60 min (15′ to 60′). Data for the right eye were compared to those for the intact, left eye and are demonstrated as percent of control. Marked reductions in the cell density of the GCL, and thickness of the IPL and INL were induced by ischemia of more than 45 min. Damage of the ONL became significant when the ischemia lasted longer than 60 min. (From Adachi et al. 1998a, with permission)

focused on the histology of the inner part of the retina for the purpose of consistent quantification.

Retinal Ischemia and Glutamate-induced Neurotoxicity

Based on our in vivo microdialysis analysis on the efflux of glutamate from the cat retina during and after ischemia, an increase in the release of glutamate occurred during ischemia, but an approximately six times larger increase in the release of glutamate took place during the reperfusion period (Fig. 6) (Adachi et al. 1998b). A similar pattern of increased release of glutamate during the period of reperfusion was reported in the rabbit retina (Louzada et al. 1992). By contrast, transient cerebral ischemia produces a spike-like release of glutamate during the period of ischemia (Benveniste et al. 1984; Globus et al. 1991; Uchiyama et al. 1994). The apparent difference in the time course of glutamate release may relate to the difference in target tissue and experimental conditions.

Neurotransmitters exert their biological effects by interacting with receptor substances in cells. On pharmacological grounds, glutamate receptors have been subdivided into five main subtypes according to their most selective agonists: NMDA, kainate, α-amino-3-hydroxy-5-methyl-4-isoxazole propionate (AMPA), 2-amino-4-phosphonobutyrate (L-AP4), and *trans*-1-amino-cyclopentane-1.3-dicarboxylate (ACPD). NMDA, kainate, and AMPA subtypes constitute

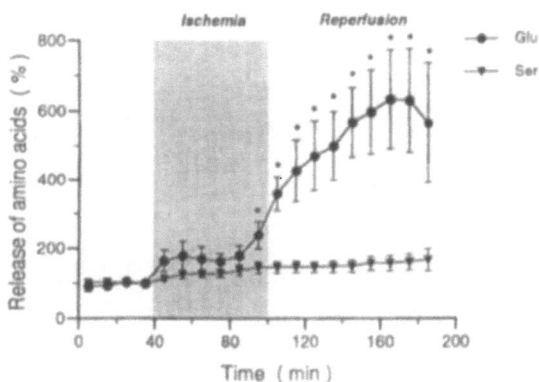

Fig. 6. Microdialysis analysis of the temporal profile of changes in the release of glutamate from cat retina subjected to 60 min of ischemia. The microdialysate level of glutamate showed a tendency to increase from the first 10 min of ischemia. A markedly larger increase in the release of glutamate occurred during the reperfusion period. Serine, a non-neurotransmitter, showed no significant changes in concentration throughout the period of ischemia and reperfusion. *$P < 0.05$ vs. serine. (From Adachi et al. 1998b, with permission)

ionotropic glutamate receptors that are linked to ion channels, whereas the other two subtypes are coupled to G-proteins with the latter widely known as metabotropic glutamate receptor whose activation results in the generation of inositol phosphate and diacylglycerol (Gasic and Hollmann 1992). Under our experimental settings, histological changes after ischemia occurred primarily in the inner part of the retina where NMDA receptors are distributed.

We first studied the effects of dizocilpine (MK-801), a selective NMDA receptor antagonist, on histological changes after ischemia. When the body temperature of the experimental animals was strictly kept at 37°C during surgery, including an ischemic session and recovery from anesthesia, pretreatment with MK-801 (3–10 mg/kg i.v.) significantly inhibited the reductions in cell density of the GCL and thickness of the INL induced by ischemia in a dose-dependent manner (Fig. 7) (Adachi et al. 1998a). By contrast, in the animal pretreated with vehicle, marked cell death occurred in the inner part of the retina. It is thus suggested that NMDA receptors play a key role in the histological changes of the inner retina induced by ischemia.

Activation of each glutamate subtype receptor with its specific agonist has been known to produce different histopathological changes in the retina. That is, kainate predominantly injured bipolar and amacrine cells (Olney et al. 1974; Schwarcz and Coyle 1977; Ingham and Morgan 1983), whereas NMDA and

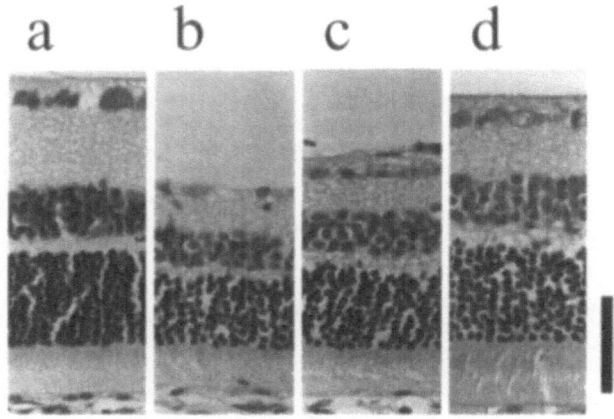

Fig. 7. Effect of intravenous injection of dizocilpine (MK-801) on histological changes after ischemia. The specimens were obtained 7 days after 45 min of ischemia. The body temperature of the experimental animal was kept at 37°C during the procedures. (a) Untreated retina. (b–d) A marked loss of cells in the GCL and a reduction in thickness of the INL occurred in the retina from an animal pretreated with vehicle (b), whereas retinas from animals pretreated with MK-801—3 mg/kg (c) and 10 mg/kg (d)—did not show these histological damages. Bar = 50 μm. (From Adachi et al. 1998a, with permission)

AMPA affected primarily amacrine cells (Gibson and Reif-Lehrer 1985; Morgan 1987; Sattayasai and Ehrlich 1987; Tung et al. 1990). As for the rat retina, intra-vitreous injection of NMDA (200 nmol) has produced histopathological changes comparable to those induced by retinal ischemia (Morizane et al. 1997). Based on our morphometric study and the results that of others, intravitreous admin-istration of NMDA was demonstrated to produce a dose-dependent reduction in the cell density of the GCL and the thickness of the IPL (Siliprandi et al. 1992; Morizane et al. 1997). These results provide a correlation between the histologi-cal distribution of glutamate receptor subtypes and vulnerability to retinal ischemia. Considering the marked protection offered by MK-801 against NMDA-induced neurotoxicity and the ischemia/reperfusion injury in the rat retina, NMDA receptors are considered to be the predominant route of gluta-mate neurotoxicity in the inner retinal injury after ischemia.

Nitric Oxide and In Vivo Experiments on Retinal Ischemia

Nitric oxide is synthesized by an enzyme called nitric oxide synthase (NOS). Three isoforms are identified: Type I, the constitutive or neuronal type, is found in neurons that are primarily amacrine cells in the retina (Osborne et al. 1993; Yamamoto et al. 1993). Type III is found in the vascular endothelium. Type II is induced by macrophages in response to some inflammation. To elucidate whether NO is involved in retinal ischemia, we studied the effects of N^{ω}-nitro-L-arginine methylester (L-NAME), a competitive NOS inhibitor of histological changes after retinal ischemia (Adachi et al. 1998a). The specimens were obtained after 7 days of reperfusion following 45 min of ischemia, with the body temperature of the animal strictly kept at 37°C. L-NAME (10 and 30 mg/kg) or nitro-D-arginine-methylester (D-NAME), an inactive enantiomer of L-NAME (30 mg/kg) was administered intravenously 30 min before ischemia. Pretreatment with L-NAME significantly inhibited the decrease in the density of GCL and the thickness of the IPL seen after ischemia in a dose-dependent manner (Fig. 8). By contrast, D-NAME was ineffective in counteracting the ischemic injury of the inner retina, suggesting that the protective action of L-NAME occurred through its inhibitory action on NOS. Similar observations were made in the rat using N^{ω}-nitro-L-arginine, another NOS inhibitor administered to the rat up to 1 h after ischemia (Geyer et al. 1995). The finding that NOS inhibition inhibited the sub-sequent histological damage in the inner retina suggests that NO is involved in the pathogenesis of ischemic injury in the retina.

In the following experiment we tried to determine if the toxic action of NO seen after ischemia was caused by stimulating the NMDA receptors; the effects of intravitreous administration of L-NAME (3 pmol to 3 nmol) on the histologi-cal changes after intravitreous injection of NMDA (200 nmol) (Morizane et al. 1997) were examined. Single intravitreous injection of NMDA produced loss of

a   b   c   d   e

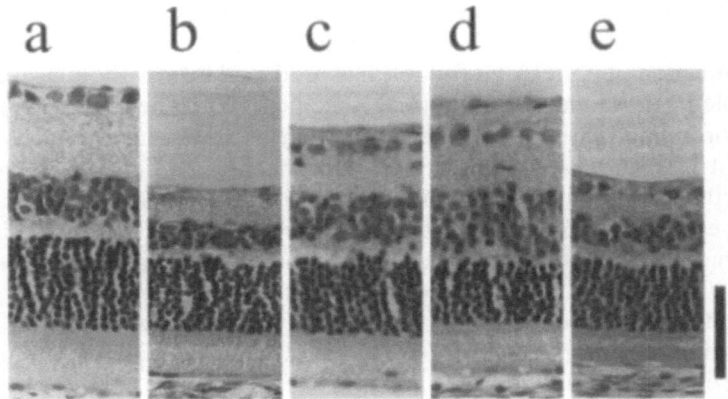

FIG. 8. Effect of intravenous injection of N^{ω}-nitro-L-arginine methylester (L-NAME) on retinal ischemia. The specimens were obtained 7 days after 45 min of ischemia. The body temperature was kept at 37°C. (**a, b**) Untreated retina (**a**) and that after ischemia (**b**) from vehicle-treated animals. (**c, d**) Pretreatment of L-NAME with 10 mg/kg (**c**) and 30 mg/kg (**d**) significantly inhibited reduction of the cell density of the GCL and thickness of the IPL in a dose-dependent manner. By contrast, nitro-D-arginine methylester (D-NAME), an inactive enantiomer, at 30 mg/kg (**e**) did not affect the ischemic injury of the inner part of the retina. Bar = 50 μm. (From Adachi et al. 1998a, with permission)

approximately half of the cells in the GCL; and there was a 70% reduction in the thickness of the INL 7 days after injection. Simultaneous injection of L-NAME with NMDA in the vitreous significantly prevented cellular loss in the GCL and reduction of the thickness of IPL induced by NMDA dose dependently (Fig. 9). Therefore, the protective action of L-NAME against retinal ischemia is believed due to the inhibition of NMDA-induced neurotoxicity. In other words, NO produced by the activation of NMDA receptors in response to ischemia plays a lethal role in the pathogenesis of an ischemic injury in the inner part of the retina. A nonselective NOS inhibitor such as L-NAME inhibits not only neuronal NOS (nNOS) from neurotoxic action but also endothelial NOS (eNOS) from vasodilating action. Despite the potential ameliorating action of NO as a vasodilator, our in vivo experiments clearly suggest a principal role of NO-induced neurotoxicity in the pathogenesis of retinal ischemia.

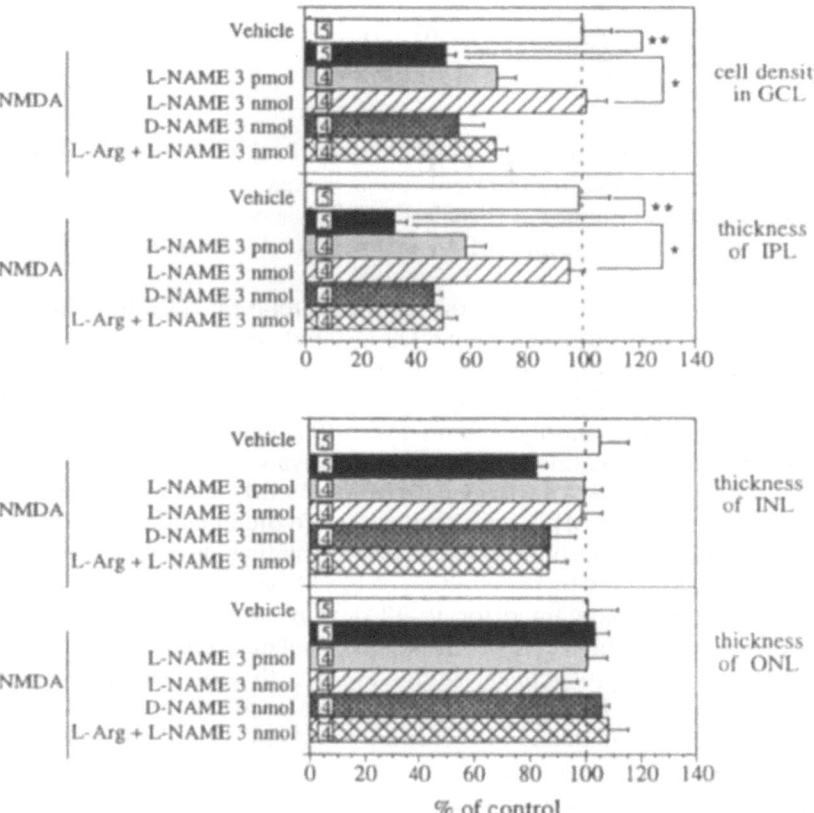

FIG. 9. Effect of intravitreous injection of L-NAME on N-methyl-D-aspartate (NMDA)-induced neurotoxicity in the retina. The specimens were obtained 7 days after intravitreous injection of NMDA or other drugs. Intravitreous injection of 200nmol NMDA produced marked cell loss in the GCL and a reduction in the thickness of the IPL. Simultaneous intravitreous injection of L-NAME (3pmol or 3nmol) with NMDA (200nmol) significantly prevented the loss of cells in the GCL and reduction of the thickness of the IPL induced by NMDA in a dose-dependent manner. Co-injection of D-NAME (3nmol) with NMDA (200nmol) into the vitreous did not affect the NMDA-induced retinal damage. The protective effect of L-NAME (3nmol) was inhibited by simultaneous injection of L-arginine (300nmol), a substrate for nitric oxide synthase (NOS). The degree of NMDA-induced retinal damage was quantified by measuring the density of cells (cells/mm) in the GCL within 1mm of the optic disk, and the thickness of the IPL, INL, and ONL about 0.5mm from the optic disk. The mean values for the right eye of each animal were normalized to those for the left (nontreated) eye and are shown as percents (abscissa). (From Morizane et al. 1997, with permission)

NO in an In Vitro Experiment on Glutamate Neurotoxicity

Although the toxic action of NO predominated in the in vivo experiments on glutamate neurotoxicity after transient retinal ischemia, our in vitro experiments using cultured retinal neurons revealed another aspect of NO: That is, retinal neurons were protected from NMDA receptor-mediated neurotoxicity by NO (Kashii et al. 1996).

Primary cultures obtained from the retinas of fetal rats (16- to 19-day gestation) were used for the experiments. Dissociated retinal cells were plated as single-cell suspensions on plastic or glass coverslips, which were then placed in Falcon 60-mm dishes (4.5×10^6 to 6×10^6 cells/dish) in a plating medium of Eagle's minimal essential medium supplemented with 10% heat-inactivated fetal bovine serum (1–9 days after plating) or 10% heat-inactivated horse serum (10–14 days after plating), glutamine (2 mM), glucose (total 11 mM), sodium bicarbonate (24 mM), and HEPES (10 mM). Retinal cultures were then maintained at 37°C in a humidified 5% CO_2 atmosphere. After 8 days of plating, the cell division of non-neuronal cells was halted by the addition of 10^{-5} M cytosine arabinoside. We used only those cultures maintained for 10–12 days in vitro. Under the conditions described above, rat retinal cultures consisted of isolated cells and clusters of cells. A previous immunocytochemical study revealed that the isolated cells were mainly amacrine cells (Kashii et al. 1994). The clusters of cells were excluded from the results, as cells embedded in the clusters could not be evaluated for cytological experiments. The neurotoxic effects of a drug on the retinal cultures were quantitatively assessed using the trypan blue exclusion method (Choi et al. 1987). Special care was taken to use Mg^{2+}-free medium for making the NMDA solution, and neither glycine nor strychnine was added to the NMDA solution based on the findings of our previous study (Kashii et al. 1994).

Immediately after a 10-min exposure to glutamate (1 mM), the cell viability remained without change. Further incubation in glutamate-free medium produced significant cell death (Fig. 10). The delayed response after exposure of cultured retinal neurons to glutamate is compatible with the delayed appearance of the ischemic injury in the inner part of the retina. A noncompetitive antagonist, MK-801, for the NMDA receptor markedly inhibited the cell death induced by glutamate, indicating that NMDA receptors are the predominant route of glutamate neurotoxicity in our culture. The effectiveness of MK-801 in terms of inhibiting cell death induced by glutamate was used as a control. Simultaneous application of either N^ω-nitro-L-arginine (N-Arg, 300 mM), an NOS inhibitor, or hemoglobin (Hb, 20 µM), an NO trapper, with glutamate (1 mM) inhibited glutamate-induced neurotoxicity (Fig. 11). These results indicate the involvement of NO in glutamate neurotoxicity.

Immediately after a 10-min exposure to sodium nitroprusside (SNP) or S-nitrosocysteine (SNOC) (500 µM), which generates NO, the cell viability did not change. Further incubation in the medium devoid of these NO donors for more

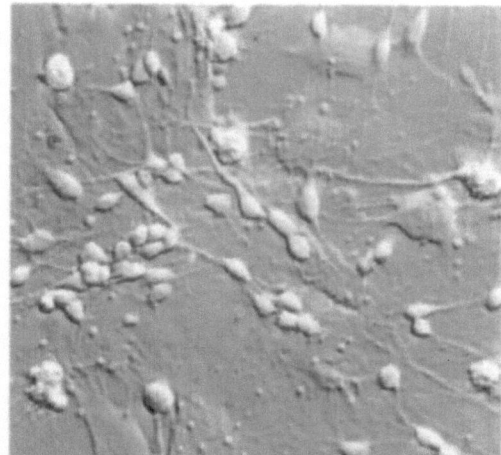

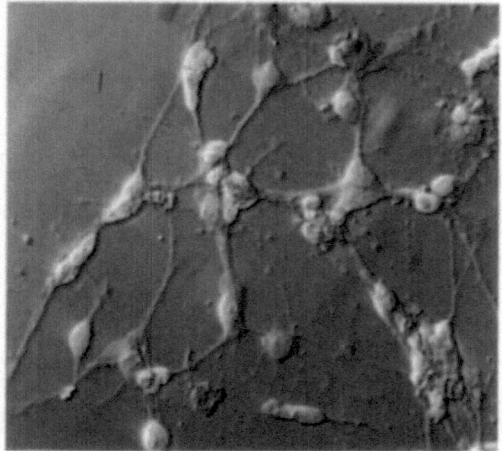

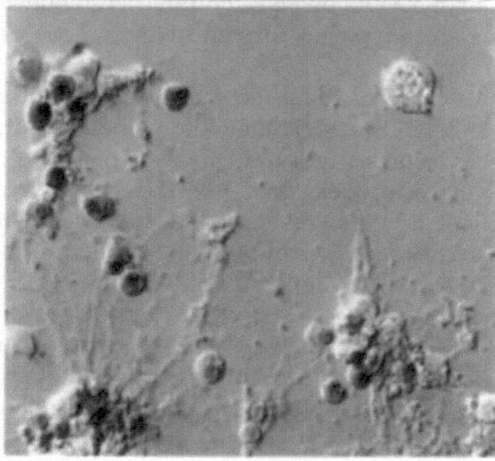

Fig. 10. Glutamate-induced neurotoxicity in the cultured retinal neurons. Photographs were taken after 1.5% trypan blue staining followed by formalin fixation using Hoffman modulation microscopy. Cells stained with trypan blue dye, which is normally excluded by the cells, were regarded as nonviable. (A) Nontreated cells (control). The cells were stained with trypan blue without application of glutamate. (B) Cells were treated with glutamate (1 mM) for 10 min. The cells were stained with trypan blue dye immediately after the 10-min exposure to glutamate. The cell viability did not change. (C) Cells were treated with glutamate for 10 min and further incubated with glutamate-free medium for 1 h. The cells were stained with trypan blue immediately after a 1-h incubation in glutamate-free medium. Marked cell death was observed after 1 h of incubation following a 10-min exposure to glutamate. Bar-50 μm. (From Kashii 1995, with permission)

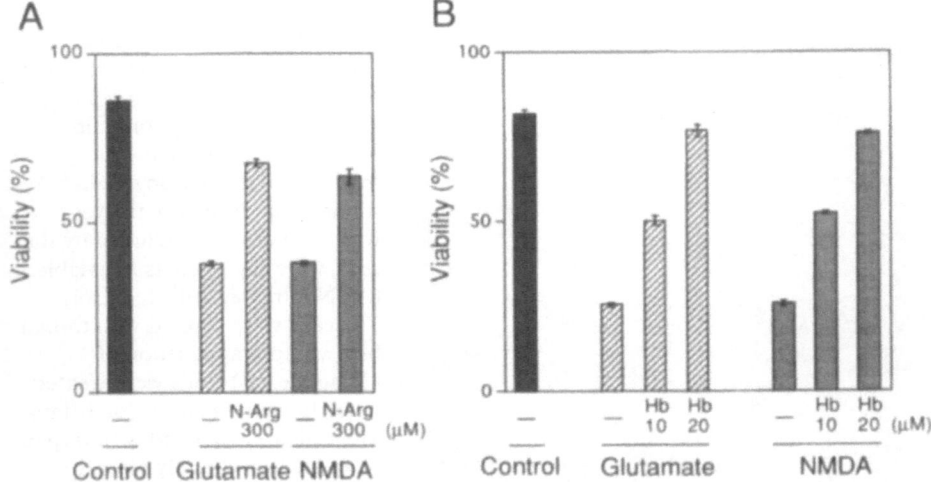

FIG. 11. The effects of N^ω-nitro-L-arginine (N-Arg), an NOS inhibitor (**A**) and hemoglobin (Hb), an NO trapper (**B**), on neurotoxicity induced by excitatory amino acids (EAAs). Cultured retinal neurons were exposed to either glutamate (1 mM) or NMDA (1 mM) for 10 min and further incubated with EAA-free medium for 1 h. The ordinate indicates the viability of cultures. Each column and bar indicate the mean and SEM of data ($n = 5$). Marked cell death induced by EAAs was inhibited by these agents in a dose-dependent manner. (From Kashii et al. 1996, with permission)

than 1 h was required to produce significant cell death. NO-induced neurotoxicity also occurred in a delayed manner (Fig. 12). A radical form of NO is known to react spontaneously with superoxide and peroxynitrite is formed as a result. Simultaneous application of superoxide dismutase (SOD), a radical scavenger that removes superoxide, with SNOC or SNP inhibited NO-induced neurotoxicity (Fig. 13), suggesting that peroxynitrite, formed by the reaction of NO with superoxide, leads to cell death.

Our electrophysiological study using a patch-clamp technique (Ujihara et al. 1993) showed that pretreatment with SNOC or SNP completely abolished the whole cell currents of cultured retinal neurons induced by the application of NMDA. The inhibition induced by NO was specific for the NMDA receptor, as pretreatment with SNP or SNOC did not affect the response induced by kainate. Neither SNOC nor SNP at 50 μM affected cell viablity when applied alone. However, simultaneous application of these NO donors at 50 μM with NMDA (1 mM) inhibited NMDA-induced neurotoxicity (Kashii et al. 1996). Thus, it is suggested that NO at low concentrations inhibits the NMDA receptors and thereby protects retinal neurons from cell death induced by glutamate. NO produced in excess, interacting with oxygen radicals, becomes toxic and mediates glutamate neurotoxicity.

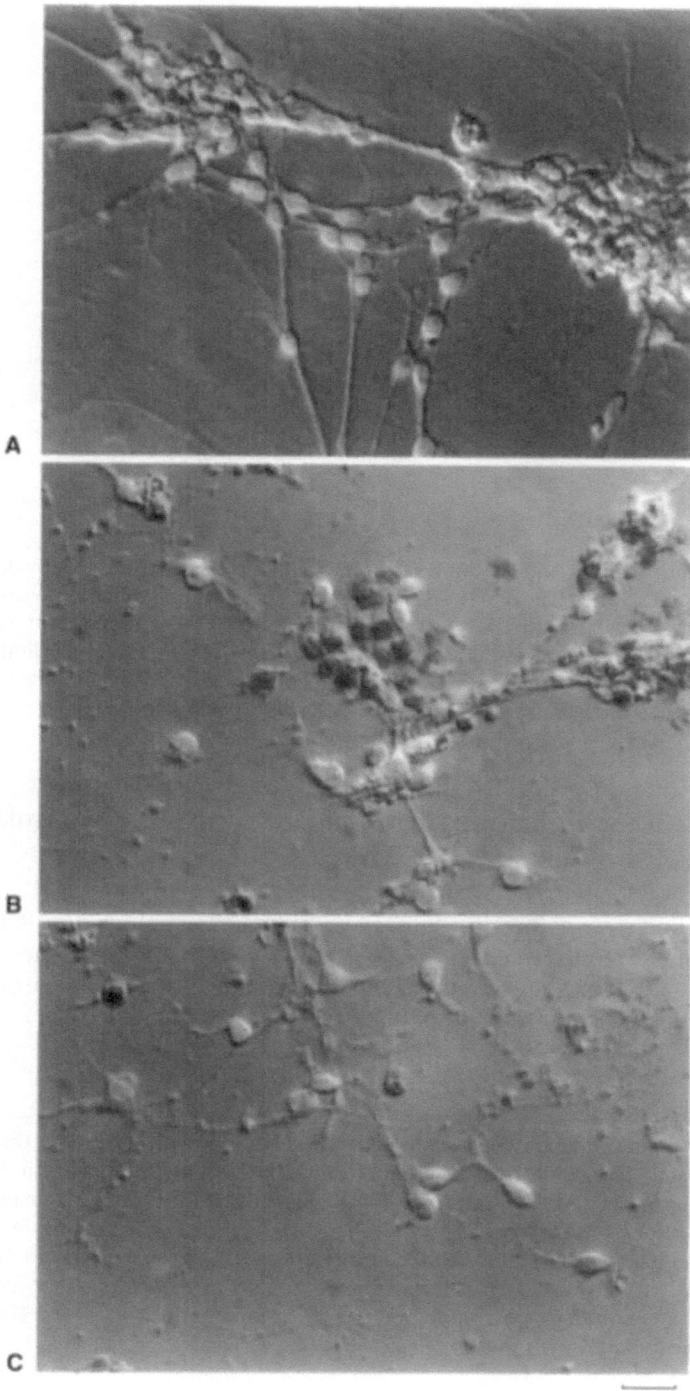

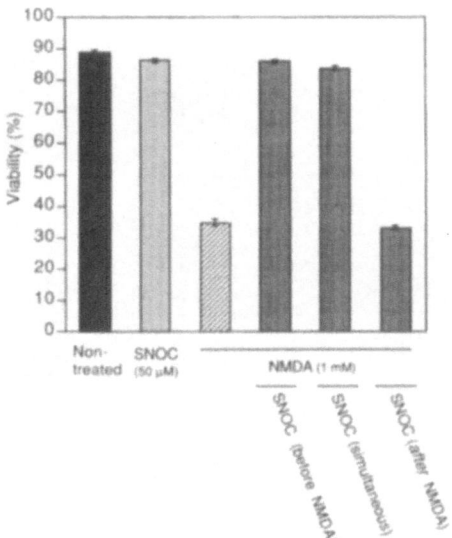

FIG. 13. Effects of superoxide dismutase (SOD) on neurotoxicity induced by NMDA and S-nitrosocysteine (*SNOC*), an NO-producing agent. Cultures were exposed to either NMDA (1 mM) or SNOC (500 μM) for 10 min followed by a 1-h incubation with normal medium. SOD (100 U/ml) was added to NMDA- or SNOC-containing medium. Simultaneous application of SOD with NMDA or SNOC inhibited cell death induced by each of the drugs applied alone. (From Kashii et al. 1996, with permission)

Pharmacological Applications of NO Pathway Against Retinal Glutamate Neurotoxicity

We have found several intrinsic substances that protect cultured retinal neurons from glutamate-induced neurotoxicity (Table 1). In this section, the site of the protective action of some of them against retinal glutamate neurotoxicity is detailed in reference to the NO pathway (Fig. 14).

◄──

FIG. 12. Effects of sodium nitroprusside (SNP), an NO donor, on cultured retinal neurons. (**A**) Nontreated cells (control). (**B**) Cells treated with SNP (500 μM) for 10 min and further incubated with SNP-free medium for 1 h. Cell death did not occur immediately after the 10-min exposure to SNP; further incubation in SNP-free medium was required to obtain significant cell death. (**C**) Cells treated with SNP (500 μM) and Hb (20 μM) simultaneously for 10 min and further incubated in SNP-free medium. SNP-induced neurotoxicity was inhibited by the simultaneous application of Hb. Bar = 50 μm (From Kashii 1995, with permission)

TABLE 1. Protective intrinsic substances

Neurotransmitters
    Dopamine (D1 receptors)
    Acetylcholine (nicotinic receptors)
Neurotrophins
    BDNF
Zinc
Calcineurin (inhibitors)
    Cyclosporin A
    FK-506
S-adenosylmethionin
    Vitamin B12 (methyl cobalamin)

BDNF, brain-derived neurotrophic factor.

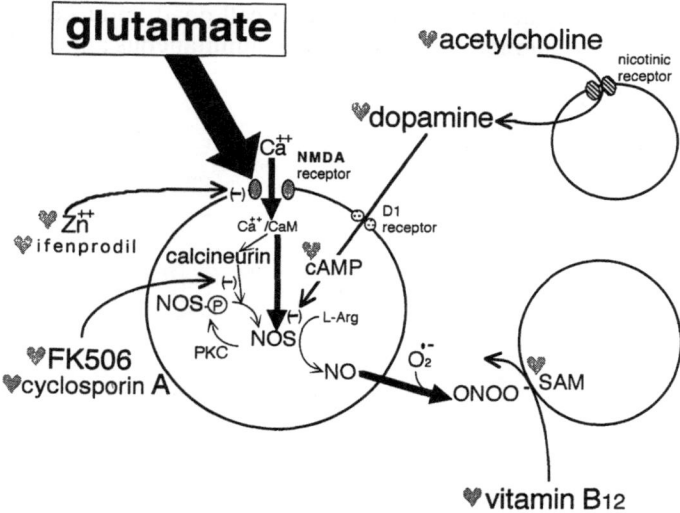

FIG. 14. Pharmacological applications of the NO pathway against retinal glutamate neurotoxicity. Dopamine protects retinal neurons against NMDA receptor-mediated glutamate neurotoxicity via D_1 receptors. Acetylcholine counteracts retinal glutamate neurotoxicity by stimulating nicotinic receptors, which release dopamine. Zn^{2+} and ifenprodil selectively bind to the affinity sites for Zn^{2+} and the polyamine site of the NMDA receptor, respectively; and their selective blockade of the NMDA receptors results in protection. FK 506 and cyclosporin A, which inhibit calcineurin, prevent the activation of NOS and its subsequent NO production. $O_2^{\cdot-}$, superoxide; $ONOO^-$, peroxynitrite; Ca^{++}/CaM, Ca^{2+}-dependent calmodulin; NOS, nitric oxide synthase; L-Arg, L-arginine; NO, nitric oxide; PKC, protein kinase C; vitamin B_{12}, methyl cobalamin; SAM, S-adenosyl methionine

Neurotransmitters and Neuromodulators

The dual actions of NO as a messenger molecule and as a toxin are similar to those of glutamate as an excitatory neurotransmitter and excitotoxin. In this context, it is important to note that dopamine (the first substance we found) had a protective action on retinal neurons against glutamate neurotoxicity in addition to its physiological action as a neuromodulator in signal transmission in the retina (Kashii et al. 1994). Dopamine exerted its protective action through the D_1 receptors by inhibiting the step in which NO was generated (Fig. 15). We proposed that chemical agents formerly believed to be transmitters or modulators might have a toxic or protective role in the life-regulatory function of retinal neurons (Kashii et al. 1994). Recently, we found that acetylcholine protected retinal neurons from glutamate neurotoxicity by way of nicotinic receptors. Acetylcholine is an excitatory neurotransmitter in the retina primarily synthesized by amacrine cells. JTC-517 is a selective agonist for a neuronal nicotinic acetylcholine receptor that consists of α4 and non-α1 subunits. This subtype of nicotinic acetylcholine receptor has been found to be prevalently distributed in the INL. Pretreatment with JTC-517 at concentrations of 0.001–1.000 µM 12 h prior to 500 µM glutamate exposure inhibited cell death induced by glutamate in a dose-dependent manner. It is of note that simultaneous application of SCH

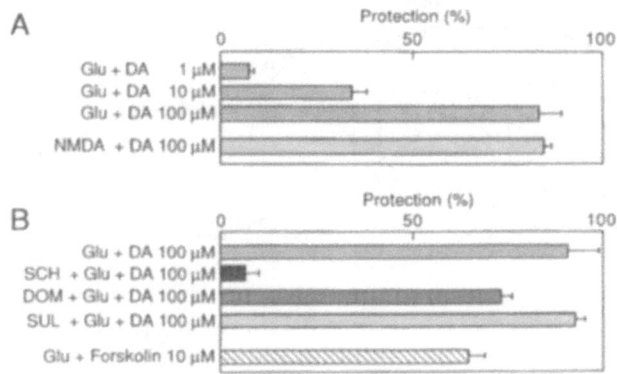

Fig. 15. Protective effects of dopamine on excitatory amino acid (EAA)-induced neurotoxicity in cultured retinal neurons. (**A**) Dose-dependent protection of dopamine (*DA*) against neurotoxicity induced by an EAA. *Glu*, glutamate (1 mM); NMDA (1 mM). (**B**) Effects of dopamine receptor blockers on the protective action of dopamine against glutamate neurotoxicity and protective effects of forskolin against EAA-induced neurotoxicity. *SCH*, SCH 23390 (5 µM); *DOM*, domperidone (5 µM); *SUL*, sulpiride (5 µM). The abcissa represents protection, which was calculated based on the following equation: Protection (%) = ([D − E]/[C − E]) × 100, where D is the viability of the cultures treated with a drug (dopamine, glutamate, NMDA), E is the viability after EAA treatment, and C is the viability of control cultures. (From Kashii et al. 1994, with permission)

23390, a dopamine D_1 receptor antagonist, with JTC-517 inhibited the protective action offered by JTC-517. Considering that nicotinic stimulation was reported to release dopamine in the retina (Myhr and McReynolds 1996), it is suggested that nicotinic stimulation induced dopamine release, which subsequently protected retinal neurons from glutamate neurotoxicity through D_1 receptors. Neurotransmitters thus seem to have some action on cell viablility in addition to their primary role as a "neuronal" signal transmitter.

Furthermore, Zn^{2+} also showed a dual action as (1) an inhibitory signaling molecule in glutamatergic synaptic transmission and (2) a protective agent against glutamate excitotoxicity. According to our patch-clamp study, $30\,\mu M$ Zn^{2+} inhibited 50% of the whole cell currents evoked by the application of NMDA ($50\,\mu M$) at potentials more positive than $-80\,mV$ (Ujihara et al. 1993). Starting at the same concentration of Zn^{2+}, neurotoxic effects of glutamate or NMDA were completely inhibited (Kikuchi et al. 1995). Zn^{2+} has been known to be released with glutamate in the retina (Wu et al. 1993) and in the brain (Assaf and Chung 1984; Howell et al. 1984; Aniksztejn et al. 1987; Frederickson 1989). The concentration of Zn^{2+} in the synaptic clefts of the central nervous system during the intense presynaptic neuronal firing is estimated at about $200–300\,\mu M$ (Assaf and Chung 1984; Howell et al. 1984; Aniksztejn et al. 1987; Frederickson 1989), which is sufficient to protect against the toxic action of glutamate, as shown in our previous study (Kikuchi et al. 1995). The retina and choroid contain the highest Zn^{2+} concentrations in ocular tissues. The concentrations of Zn^{2+} in most human organs ranged between 20 and $30\,\mu g/g$ dry weight. By contrast, those in human retina and choroid have been calculated at means of 464 and $472\,\mu g/g$ dry weight, respectively; these levels are extremely high in comparison with those in other soft tissues in humans and experimental animals (Karcioglu 1982). An unusually high concentration of Zn^{2+} in the retina is believed to contribute to the stabilization of postsynaptic neurons with glutamate receptors. Thus, chemical agents such as neurotransmitters or modulators are thought to have some action on cell viability (either good or bad) in addition to their primary role as a "neuronal" signal transmitter.

Calcineurin

Calcineurin, the major Ca^{2+}/calmodulin-regulated protein phosphatase, is enriched in neuronal tissue including the retina. It co-localizes with the immunophilins cyclophilin and FK-binding protein (FKBP). The immunosuppressant drugs cyclosporin A and FK506, which form drug–immunophilin complexes, are highly specific inhibitors of calcineurin.

The administration of FK506 or cyclosporin A for 10 min prior to glutamate ($500\,\mu M$) application and during the 10-min exposure to glutamate inhibited cell death in a dose-dependent manner at 0.1 nM to $1.0\,\mu M$ (Kikuchi et al. 1998). Rapamycin is a structural analog of FK506 and a high-affinity ligand for FKBP, but it has no effect on calcineurin activity (Abraham and Wiederrecht 1996). The protective effect of FK506 was reversed by simultaneous application of

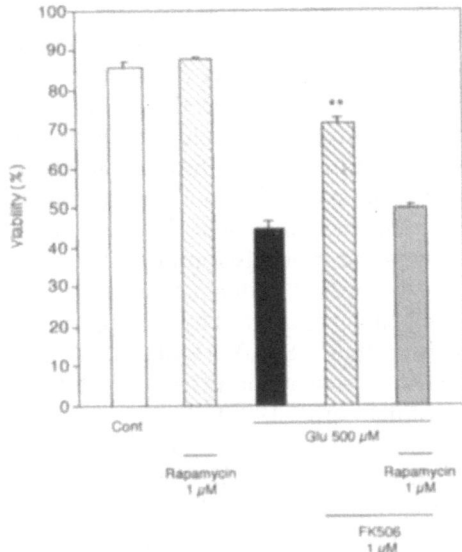

F<small>IG</small>. 16. Effect of rapamycin, a competitive inhibitor of FK-506 for binding FK-506-binding protein, on the protective action of FK-506 against glutamate neurotoxicity. Simultaneous application of rapamycin with FK506 did not inhibit cell death induced by glutamate (*Glu*), whereas it had no effect on cell viability when applied alone. **$P < 0.01$ vs. the *black column*. (From Kikuchi et al. 1998, with permission)

rapamycin, indicating that the protective action of FK506 against glutamate neurotoxicity is mediated by the inhibition of calcineurin (Fig. 16).

Application of FK506 did not affect the intracellular elevation of Ca^{2+} in response to glutamate exposure, indicating that FK506 had no direct effects on the glutamate receptors. Furthermore, FK506 inhibited cell death induced by increasing the intracellular Ca^{2+} concentration with the application of A23187, a Ca^{2+} ionophore, but it did not inhibit cell death induced by an NO-producing agent, SNOC. Therefore, the protective effects of FK506 was considered to take place before NO production and downstream of the Ca^{2+} influx. This notion was further supported by the subsequent measurements of NOS activity, which demonstrated that pretreatment of cultured retinal neurons with FK506 abolished the NOS activity evoked by glutamate application. Thus, FK 506 was considered to protect cultured retinal neurons from glutamate neurotoxicity by inhibiting calcineurin, which prevented activation of NOS and its subsequent generation of NO.

Vitamin B_{12}

According to our study of the effects of water-soluble B vitamins on retinal glutamate neurotoxicity, glutamate neurotoxicity was inhibited in the cultures

TABLE 2. Effects of chronic administration of B vitamins on glutamate-induced neurotoxicity

Vitamin	Effect
Thiamine (10–100 μM)	Protective
Thiamine pyrophosphate (50 μM)	Protective
Pyridoxine (20–200 μM)	Protective
Pyridoxal phosphate (200 μM)	Protective
Nicotinamide (200 μM)	Protective
Riboflavin (1–10 μM)	None
Pantothenate (1–100 μM)	None
Biotin (0.2–20 μM)	None
Folic acid (10–100 μM)	None
Inositol (20–200 μM)	None

maintained in incubation media containing either thiamine pyrophosphate, pyridoxisal phosphate, methylcobalamin, or nicotinamide prior to glutamate exposure (Table 2) (Kaneda et al. 1997; Kikuchi et al. 1997). The site of the protective action of methylcobalamin is unique in that its chronic application inhibited cell death induced by NO. The sites of protective action of most of the substances we had tested for glutamate neurotoxicity were situated upstream of NO generation (Fig. 14). That is, once NO was synthesized, these substances were no longer effective in protecting the retinal neurons. Methylcobalamin and its intracellular metabolite S-adenosyl methionine (SAM), are only two exceptions that can counteract NO-induced neurotoxicity.

The SAM is synthesized from methionine and adenosine triphosphate by the enzyme methionine adenosyl transferase in many mammalian tissues including the retina. The daily dietary intake of methionine is not sufficient to provide the total amount required for SAM synthesis (Bottigllieri et al. 1994), so methionine must be regenerated from homocysteine by vitamin B_{12}-dependent methionine synthase. Mammals are unable to synthesize homocysteine de novo. Here, the transfer reaction of a methyl group from 5-methyl tetrahydrofolate to homocysteine with regeneration of methionine, catalyzed by methionine synthase, requires methylcobalamin, an active coenzyme of the vitamin B_{12} analogs, as cofactor.

Retinal cultures treated with methylcobalamin or SAM throughout the incubation period of 10–12 days (i.e., immediately after cell plating and immediately before glutamate exposure) became tolerant not only to glutamate-induced neurotoxicity but NO-induced neurotoxicity in a dose-dependent manner (Kaneda et al. 1997). These are the only two substances that are still effective in counteracting the direct toxic effect of NO. In contrast to the chronic administration, either methylcobalamin or SAM, when administered during both the 10-min glutamate exposure and the 1-h postincubation period, did not inhibit cell death induced by glutamate.

Protective effects observed only after chronic application of methylcobalamin or SAM indicates that direct interaction of each substance with the glutamate receptors was less likely. Moreover, the intracellular Ca^{2+} recordings of the cultured retinal neurons demonstrated that chronic application of methylcobalamin or SAM did not affect the glutamate-induced Ca^{2+} influx compared with that in nontreated cells, indicating that chronic application of each drug did not alter the expression of NMDA receptors of the retinal neurons.

The membranes of the retinal neurons rich in polyunsaturated fatty acids are particularly vulnerable to free radical attack. NO alone had no toxic effects on our cultured retinal neurons. It was peroxynitrite formed by the reaction of NO with superoxide anion (O_2^-) that was thought to cause cell death (Kashii et al. 1996). The peroxynitrite readily decomposes to the hydroxy radical ($·OH$) (Beckman et al. 1990), which easily removes hydrogen atoms of the double bonds of polyunsaturated fatty acid side chains in the membrane of retinal neurons. The action of a single $·OH$ provokes a subsequent chain reaction, generating numerous toxic reactants that rigidify membranes by cross-linking and disrupt membrane integrity (Coyle and Puttfarcken 1993). SAM has been known to increase the fluidity of neuronal membranes by increasing the phosphatidyl-choline content of the lipid core of the membrane (Fig. 17) (Muccioli et al. 1992). Therefore, these SAM-mediated methylations are believed to alter the membrane properties of cultured retinal neurons that reinforce the cell body, making it more resistant to radical toxicity.

Conclusion

Based on our in vivo and in vitro experiments described here, it becomes certain that NO, a simple gas with a free radical, can act as a signaling molecule and a neurotoxin in glutamate neurotoxicity, which is responsible for the pathogenesis of the ischemic injury of the inner part of the retina. Some issues must still be addressed, however.

It has been puzzling how cell death occurs specifically in cells with NMDA receptors; and NO, once formed in a cell, rapidly diffuses out into neighboring cells as a messenger molecule that is not toxic by itself. Formation of peroxynitrite, which leads to cell death, requires strict co-localization of NO and O_2^- production. The site of injury is thus determined by the site where NO meets O_2^-. The availability and distribution of O_2^- determines which cells die. When the substrate concentration of NOS (i.e., L-arginine) is low and rate-limiting, the electron transfer to the hem in the NOS becomes uncoupled from NO production, and the NOS then produces a mixture of O_2^- and NO (Heinzel et al. 1992; Pou et al. 1992). The uncoupled NOS reaction is suggested to result in subsequent neurotoxicity in the presence of brain ischemia. However, the population of amacrine cells that possesses NOS is too small to explain the massive cell loss that occurs in the retina after an ischemic insult. Although it was in cultured cerebellar granule cells that there was more resistance to NO-induced neurotoxicity and in which peroxinitrite played little role in its neurotoxicity, O_2^- has been demonstrated to be generated upon NMDA receptor stimulation

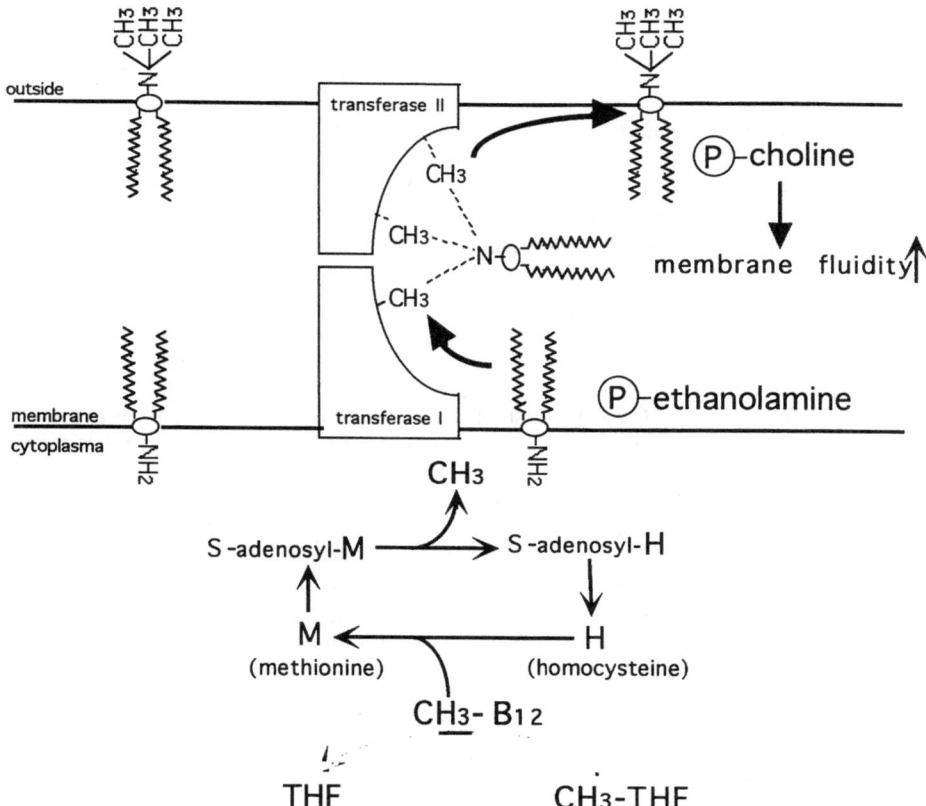

Fig. 17. SAM increases the membrane fluidity by increasing the phosphatidylcholine content of the lipid core of the membrane. *transferase I* and *II*, phospholipid *N*-methyltransferase; ℗-*ethanolamine*, phosphatidyl ethanolamine; ℗-*choline*, phosphatidyl choline; *CH₃-B₁₂*, methyl cobalamin; *THF*, tetrahydrofolate

(Lafon-Cazal et al. 1993; Fagni et al. 1994). Arachidonic acid produced by activation of phospholipase A_2 in response to elevated intracellular $[Ca^{2+}]$, which is induced by NMDA receptor stimulation, as is suggested to occur in the cerebellar granule cells, could be one candidate for such sources of O_2^- in the retina.

The NO-induced neurotoxicity does not occur immediately after exposure to NO but takes place in a delayed fashion (Fig. 12). A period of incubation in normal medium is necessary for cell death to occur. What is responsible for the delayed onset? There is still much to be done to elucidate the precise mechanism of NO neurotoxicity in the retina.

References

Abraham R, Wiederrecht G (1996) Immunopharmacology of rapamycin. Annu Rev Immunol 14:483–510

Adachi K, Fujita Y, Morizane C, Akaike A, Ueda M, Satoh M, Masai H, Kashii S, Honda Y (1998a) Inhibition of N-methyl-D-aspartate receptors and nitric oxide synthase reduces ischemic injury of the retina. Eur J Pharmacol 350:53–57

Adachi K, Kashii S, Masai H, Ueda M, Morizane C, Kaneda K, Kume T, Akaike A, Honda Y (1998b) Mechanism of the pathogenesis of glutamate neurotoxicity in retinal ischemia. Graefes Arch Clin Exp Ophthalmol 236:766–774

Aniksztejn L, Charton G, Ben AY (1987) Selective release of endogenous zinc from the hippocampal mossy fibers in situ. Brain Res 404:58–64

Assaf SY, Chung SH (1984) Release of endogenous Zn^{2+} from brain tissue during activity. Nature 308:734–736

Beckman JS, Beckman TW, Chen J, Marshall PA, Freeman BA (1990) Apparent hydroxyl radical production by peroxynitrite: implications for endothelial injury from nitric oxide and superoxide. Proc Natl Acad Sci USA 87:1620–1624

Benveniste H, Drejer J, Shousboe A, Diemer NH (1984) Elevation of the extracellular concentrations of glutamate and aspartate in the rat hippocampus during transient cerebral ischemia monitored by intracerebral microdialysis. J Neurochem 43:1369–1374

Bottigllieri T, Hyland K, Reynolds EH (1994) The clinical potential of ademethionine (S-adenosylmethionine) in neurological disorders. Drugs 48:137–152

Brown CG, Magargal LE, Sergott R (1986) Obstruction of the retinal and choroidal circulation. Ophthalmology 93:1373–1382

Choi DW, Maulucci-Gedde M, Kriegstein AR (1987) Glutamate neurotoxicity in cortical cell culture. J Neurosci 7:357–368

Coyle JT, Puttfarcken P (1993) Oxidative stress, glutamate and neurodegenerative disorders. Science 269:689–695

Faberowski N, Stefansson E, Davidson RC (1989) Local hypothermia protects the retina from ischemia. Invest Ophthalmol Vis Sci 30:2309–2313

Fagni L, Lafon-Cazal M, Rondouin G, Manzoni O, Lerner-Natoli M, Bockaert J (1994) The role of free radicals in NMDA-dependent neurotoxicity. Prog Brain Res 103:381–390

Frederickson CJ (1989) Neurobiology of zinc and zinc-containing neurons. Int Rev Neurobiol 31:145–238

Gasic GP, Hollmann M (1992) Molecular neurobiology of glutamate receptors. Annu Rev Physiol 54:507–536

Geyer O, Almog J, Lupu-Meiri M, Lazar M, Oron Y (1995) Nitric oxide synthase inhibitors protect rat retina against ischemic injury. FEBS Lett 374:399–402

Gibson B, Reif-Lehrer L (1985) Mg^{2+} reduced N-methyl-D-aspartate neurotoxicity in embryonic chick neural retina in vitro. Neurosci Lett 57:13–17

Globus MYM, Busto R, Martinez E, Ginsberg MD (1991) Comparative effect of transient global ischemia on extracellular levels of glutamate, glycine and γ-aminobutyric acid in vulnerable and nonvulnerable brain regions in the rat. J Neurochem 57:470–478

Harris A, Sergott RC, Spaeth GL, Katz JL, Shoemaker JA, Martin BJ (1994) Color doppler analysis of ocular vessel blood velocity in normal tension glaucoma. Am J Ophthalmol 118:642–649

Hayreh SS (1971) Pathogenesis of occlusion of the central retinal vessels. Am J Ophthalmol 72:998–1011

Hedges TR (1962) Ophthalmoscopic findings in internal cartotid occlusion. Bull Johns Hopkins Hosp 3:89–97

Heinzel B, John M, Klatt P, Böohme E, Mayer B (1992) Ca^{2+}/calmodulin-dependent formation of hydrogen peroxide by brain nitric oxide synthase. Biochem J 281:627–630

Howell GA, Welch MG, Frederickson CJ (1984) Stimulation-induced uptake and release of zinc in hippocampal slices. Nature 308:736–738

Ingham CA, Morgan IG (1983) Dose-dependent effects of intravitreal kainic acid on specific cell types in chicken retina. Neuroscience 9:165–181

Kaneda K, Kikuchi M, Kashii S, Honda Y, Maeda T, Kaneko S, Akaike A (1997) Effects of B vitamins on glutamate-induced neurotoxicity in retinal cultures. Eur J Pharmacol 322:259–264

Karcioglu ZA (1982) Zinc in the eye. Surv Ophthalmol 27:114–122

Kashii S (1995) The role of nitric oxide in the ischemic retina. J Jpn Ophthalmol Soc 99:1361–1376

Kashii S, Mandai M, Kikuchi M, Honda Y, Tamura Y, Kaneda K, Akaike A (1996) Dual actions of nitric oxide in N-methyl-D-aspartate receptor-mediated neurotoxicity in cultured retinal neurons. Brain Res 711:93–101

Kashii S, Takahashi M, Mandai M, Shimizu H, Honda Y, Sasa M, Ujihara H, Tamura Y, Yokota T, Akaike A (1994) Protective action of dopamine against glutamate neurotoxicity in the retina. Invest Ophthalmol Vis Sci 35:685–695

Kearns TP (1983) Differential diagnosis of central retinal vein obstruction. Ophthalmology 90:475–480

Kearns TP, Hollenhorst RW (1963) Venous-stasis retinopathy of occlusive disease of the carotid artery. Mayo Clin Proc 38:304–312

Kikuchi M, Kashii S, Honda Y, Tamura Y, Kaneda K, Akaike A (1997) Protective effects of methylcobalamin, a vitamin B$_{12}$ analog, against glutamate-induced neurotoxicity in retinal cell culture. Invest Ophthalmol Vis Sci 38:848–854

Kikuchi M, Kashii S, Honda Y, Ujihara H, Sasa M, Tamura Y, Akaike A (1995) Protective action of zinc against glutamate neurotoxicity in cultured retinal neurons. Invest Ophthalmol Vis Sci 36:2048–2053

Kikuchi M, Kashii S, Mandai M, Yasuyoshi H, Honda Y, Kaneda K, Akaike A (1998) Protective effects of FK506 against glutamate-induced neurotoxicity in retinal cell culture. Invest Ophthalmol Vis Sci 39:1227–1232

Lafon-Cazal M, Pietri S, Culcasi M, Bockaert J (1993) NMDA-dependent superoxide production and neurotoxicity. Nature 364:535–537

Louzàda JP, Dias JJ, Santos WF, Lachat JJ, Bradford HF, Coutinho NJ (1992) Glutamate release in experimental ischemia of the retina: an approach using microdialysis. J Neurochem 59:358–363

Miller NR (1991) Walsh and Hoyt's clinical neuro-ophthalmology (4th ed). Williams & Wilkins, Baltimore, pp 2336–2344

Morgan IG (1987) AMPA is a powerful neurotoxin in the chicken retina. Neurosci Lett 79:267–271

Morizane C, Adachi K, Furutani I, Fujita Y, Akaike A, Kashii S, Honda Y (1997) N^{ω}-nitro-L-arginine methyl ester protects retinal neurons against N-methyl-D-aspartate-induced neurotoxicity in vivo. Eur J Pharmacol 328:45–49

Muccioli G, Scordamaglia A, Bertacco S (1992) Effect of S-adenosyl-L-methionine on brain muscarinic receptors of aged rats. Eur J Pharmacol 227:293–299

Myhr KR, McReynolds JS (1996) Cholinergic modulation of dopamine release and horizontal cell coupling in mudpuppy retina. Vision Res 36:3933–3938

Nicolela MT, Drance SM, Rankin SJA, Buckley AR, Walman BE (1996) Color doppler imaging in patients with asymmetric glaucoma and unilateral visual field loss. Am J Ophthalmol 121:502–510

Olney JW, Rhee V, Ho OL (1974) Kainic acid: a powerful neurotoxic analogue of glutamate. Brain Res 77:507–512

Osborne NN, Barnett NL, Herrera AJ (1993) NADPH diaphorase localization and nitric oxide synthetase activity in the retina and anterior uvea of the rabbit eye. Brain Res 610:194–198

Pou S, Pou WS, Bredt DS, Snyder SH, Rosen GM (1992) Generation of superoxide by purified brain nitric oxide synthase. J Biol Chem 267:24173–2416

Rankin SJA, Walman BE, Buckley AR, Drance SM (1995) Color doppler imaging and spectral analysis of the optic nerve vasculature in glaucoma. Am J Ophthalmol 119:685–693

Rojanapongpun P, Drance SM, Morrison BJ (1993) Ophthalmic artery flow velocity in glaucomatous and normal subjects. Br J Ophthalmol 77:25–29

Ross Russell RW, Page NGR (1983) Critical perfusion of brain and retina. Brain 106:419–434

Sattayasai J, Ehrlich D (1987) Morphology of quisqualate-induced neurotoxicity in the chicken retina. Invest Ophthalmol Vis Sci 28:106–117

Schwarcz R, Coyle JT (1977) Kainic acid: neurotoxic efects after intraocular injection. Invest Ophthalmol 16:141–148

Siliprandi R, Canella R, Carmignoto G, Schiavo N, Zenellato A, Zanoni R, Vantini G (1992) N-Methyl-D-aspartate-induced neurotoxicity in the adult rat retina. Vis Neurosci 8:567–573

Tung NN, Morgan IG, Ehrlich D (1990) A quantitative analysis of the effects of excitatory neurotoxins on retinal ganglion cells in the chick. Vis Neurosci 4:217–223

Uchiyama TY, Araki H, Tae T, Otomo S (1994) Changes in the extracellular concentrations of amino acids in the rat striatum during transient focal cerebral ischemia. J Neurochem 62:1074–1078

Ujihara H, Akaike A, Tamura Y, Yokota Y, Sasa M, Kashii S, Honda Y (1993) Blockade of retinal NMDA receptors by sodium nitroprusside is probably due to nitric oxide formation. Jpn J Pharmacol 61:375–377

Wu SM, Qiao X, Noebels JL, Yang XL (1993) Localization and modulatory actions of Zn^{2+} in vertebrate retina. Vision Res 33:2611–2616

Yamamoto R, Bredt DS, Snyder SH, Stone RA (1993) The localization of nitric oxide synthase in the rat eye and related cranial ganglia. Neuroscience 54:189–200

Subject Index